Katharina Dörr

Travel Time Models and Throughput Analysis of Dual Load Handling Automated Storage and Retrieval Systems in Double Deep Storage

AF201494

WISSENSCHAFTLICHE BERICHTE

Institut für Fördertechnik und Logistiksysteme
am Karlsruher Institut für Technologie (KIT)

BAND 92

Travel Time Models and Throughput Analysis of Dual Load Handling Automated Storage and Retrieval Systems in Double Deep Storage

by
Katharina Dörr

KIT Scientific Publishing

Dissertation, Karlsruher Institut für Technologie
KIT-Fakultät für Maschinenbau

Tag der mündlichen Prüfung: 12. März 2018
Referenten: Prof. Dr.-Ing. Kai Furmans, Prof. Dr.-Ing. Johannes Fottner

Impressum

KIT Scientific Publishing

Karlsruher Institut für Technologie (KIT)
KIT Scientific Publishing
Straße am Forum 2
D-76131 Karlsruhe

KIT Scientific Publishing is a registered trademark
of Karlsruhe Institute of Technology.
Reprint using the book cover is not allowed.

www.ksp.kit.edu

This document – excluding the cover, pictures and graphs – is licensed under a Creative Commons Attribution-Share Alike 4.0 International License (CC BY-SA 4.0): https://creativecommons.org/licenses/by-sa/4.0/deed.en

The cover page is licensed under a Creative Commons Attribution-No Derivatives 4.0 International License (CC BY-ND 4.0): https://creativecommons.org/licenses/by-nd/4.0/deed.en

Print on Demand 2018 – Gedruckt auf FSC-zertifiziertem Papier

ISSN 0171-2772
ISBN 978-3-7315-0793-2
DOI 10.5445/KSP/1000082200

Travel Time Models and Throughput Analysis of Dual Load Handling Automated Storage and Retrieval Systems in Double Deep Storage

Zur Erlangung des akademischen Grades eines

Doktors der Ingenieurwissenschaften (Dr.-Ing.)

bei der KIT-Fakultät für Maschinenbau des
Karlsruher Institut für Technologie (KIT)
genehmigte

Dissertation

von

M.Sc. Wi.-Ing. Katharina Alissa Dörr

aus Püttlingen

Tag der mündlichen Prüfung:	12.03.2018
Hauptreferent:	Prof. Dr.-Ing. Kai Furmans
Korreferent:	Prof. Dr.-Ing. Johannes Fottner

Danksagung

Die vorliegende Arbeit entstand während meiner Tätigkeit als wissenschaftliche Mitarbeiterin am Institut für Fördertechnik und Logistiksysteme des Karlsruher Institut für Technologie.

Ich bedanke mich bei meinem Doktorvater Prof. Dr.-Ing. Kai Furmans für die Betreuung meiner Dissertation. Er hat mir viele Freiheiten gewährt und mir stets großes Vertrauen geschenkt. Prof. Dr.-Ing. Johannes Fottner danke ich für die Übernahme des Koreferats. Bei Herrn Prof. Dr. rer. nat. Frank Gauterin bedanke ich mich für die Übernahme des Pürfungsvorsitzes meiner Promotionsprüfung.

Allen aktiven und ehemaligen Kollegen des IFL danke ich für die angenehme und inspirierende Arbeitsatmosphäre, die das Erstellen einer solchen Arbeit erleichtert hat. Insbesondere danke ich Marion, Zäzilia und Andreas, die mich immer wieder aufgebaut und unterstützt haben, sowie Melanie und Holger für die kritische Korrektur meiner Arbeit. Außerdem danke ich allen Abschlussarbeitern und Hiwis, die mit viel Engagement das Thema im Kleinen vorangetrieben haben. Ein großer Dank gilt all meinen wunderbaren Freunden, die immer an mich geglaubt haben. Ihr seid die Besten.

Meiner Schwester Laura danke ich von ganzem Herzen für stetiges Ermutigen und Zureden sowie für die zahlreichen Korrekturrunden. Ganz besonderer Dank gilt meinen Eltern, da sie mir meine Ausbildung ermöglicht haben. Mama, Papa, Laura - ihr habt mich immer unterstützt und begleitet und zu dem Menschen gemacht, der ich bin. Dafür bin ich sehr dankbar.

Mein allergrößter Dank gilt Christian für unermüdliches Aufbauen, Ertragen, Zuhören, Kraftgeben und insbesondere dafür, niemals Zweifel an mir zuzulassen. Danke, dass Du da bist.

Karlsruhe, März 2018 Katharina Dörr

Kurzfassung

Vollautomatische Regalanlagen sind essentielle technische Bestandteile in Lager und Distributionszentren. Da diese Anlagen langfristige Investitionen darstellen, die bei Fehlplanung zu hohen Folgekosten führen können, sind deren Anwender bei der Auswahl und Dimensionierung auf verlässliche Ergebnisse aus der Forschung angewiesen. Solche Systeme werden kontinuierlich von deren Herstellern der Industrie weiterentwickelt, was dazu führt, dass die Praxis der wissenschaftlichen Betrachtung oftmals voraus ist.

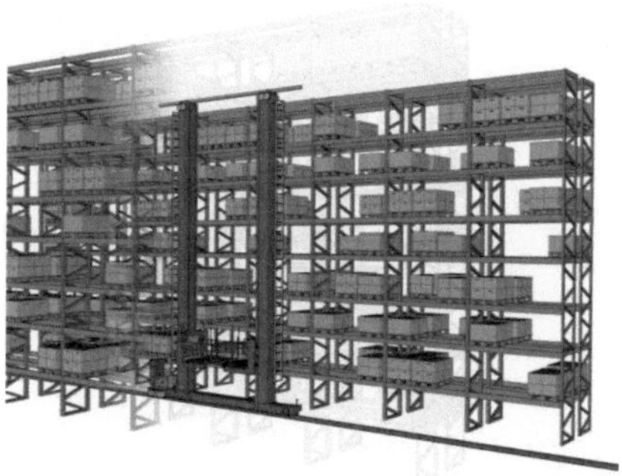

Beispiel eines vollautomatischen Lagers mit doppeltiefen Lagerplätzen und doppelter Lastaufnahme

Eine Möglichkeit die Effizienz vollautomatischer Regalanlagen zu erhöhen, ist es, die Regale mit doppeltiefen Lagerplätzen auszustatten und ein Regalbediengerät mit zweifachem Lastaufnahmemittel zu verwenden, wie in der Abbildung dargestellt. Diese Variante ermöglicht sowohl eine verbesserte Flächennutzung, als auch einen höheren Durchsatz.

Obwohl dieser Aufbau häufig in der Praxis anzutreffen ist, fehlt es bisher an genauen analytischen Formulierungen, sowie einer Betrachtung anspruchsvoller Betriebsstrategien, die den Durchsatz steigern können. Das Ziel dieser Arbeit ist es, diese Forschungslücke in zwei Schrtitten zu schließen:

Zuerst wird ein grundlegendes analytisches Spielzeitmodell für ein Vierfachspiel bei **doppeltiefer Lagerung** und dem Einsatz eines **doppelten Lastaufnahmemittels** formuliert, wobei die Wahrscheinlichkeit für eine der beiden Reihenfolgen von Einlagerungen und Auslagerungen einen Paramter des Modells darstellt. Das Modell wird schließlich mit Hilfe einer Simulation validiert.

Im zweiten Schritt werden Betriebsstrategien, die den Durchsatz im Vergleich zur zufälligen Abarbeitung erhöhen sollen, zusammengestellt. Für ausgewählte Strategien werden zudem analytische Ausdrücke hergeleitet. Mit Hilfe der mathematischen Formeln wird der Einfluss verschiedener Parameter auf die Spielzeit der Strategien untersucht. Um die Strategien im Rahmen von realistischeren Szenarien zu bewerten, wird ein Simulationsmodell verwendet, wodurch ein Vergleich dieser unter verschiedenen Einsatzbedingungen ermöglicht wird. Zum Schluss werden die Ergebnisse der Strategiebewertung diskutiert und Implikationen für deren Anwendung hinsichtlich unterschiedlicher Rahmenbedingungen abgeleitet.

Abstract

Automated storage and retrieval systems (AS/RSs) are an essential piece of warehouse technology. As they represent long-term investments, misplanning of these systems results in follow-up costs, which is why practitioners are in urgent need of sound research results for selection and sizing of these systems. The efficiency of AS/RSs is continuously improved by practitioners in industry with the consequence of industrial progress overtaking theoretical research.

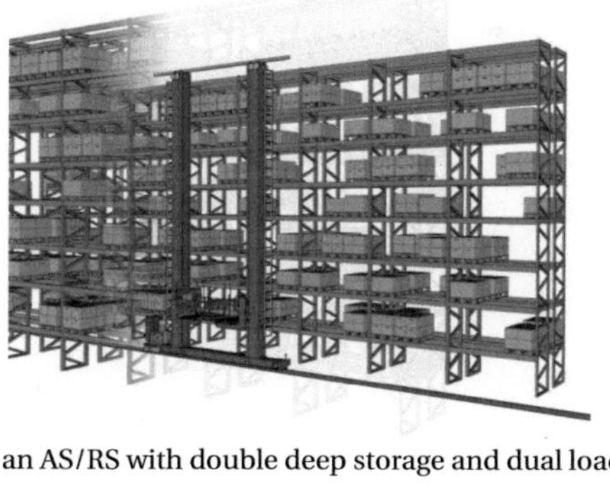

Example of an AS/RS with double deep storage and dual load handling

One possibility to increase the efficiency of automated storage and retrieval systems is to install a double deep rack structure and the usage of a storage and retrieval machine with two load handling devices as shown in the illustration. This set-up provides enhanced space utilization and increased throughput potential.

Although such systems are installed in practice, we encounter the absence of feasible analytical formulations as well as an investigation of sophisticated operating strategies to improve throughput. The objective of this thesis is to close this gap in two steps:

We formulate a general analytical travel time model for the quadruple command cycle **in double deep storage systems** with a **dual capacity load handling device**. Using a parameter that describes the execution order of a quadruple command cycle, all possible execution orders are included. The validity of the approach is shown by means of a simulation model.

We compose various routing and sequencing strategies that aim on improving throughput compared to the random execution. For selected strategies, analytical formulations are derived. In doing so, we provide basic components to describe many additional strategies in a mathematical way. To assess the strategies in consideration of real-world cases, a simulation model is used to compare strategies for various configurations and settings. Finally, we discuss results of our simulation studies as well as implications concerning implementation and application of the strategies in different environments.

Contents

1 Introduction

Confidence
-M. Kiehn, née Becker

As the global volume of trade and flow of goods is continuously increasing, cross-docking and warehousing is more important than ever. Within these logistical hubs, efficient technology is a crucial element to guarantee both supply chain performance and reliability. Automated storage and retrieval systems (AS/RSs), representing the combination of storage equipment and controls to automatically store and retrieve goods, are one of the key elements used in this context.

However, automated storage and retrieval systems entail significant investments. Besides, they are quite inflexible in terms of physical changes, once they are installed (Bartholdi and Hackman 2016, p. 200). Therefore, detailed knowledge about the performance of an AS/RS is needed to capture full economic benefit. An exact evaluation of its throughput is required to meet external requirements (i.e. customer demands) on the one hand, and to ensure a smooth operation among connected processes as found in distribution centers on the other hand. Reliable planning of automated storage and retrieval systems prevents follow-up costs, while solid evaluation of changing external conditions ensures timely modifications of these rather inflexible systems. Both reasons emphasize why practitioners are in urgent need of sound research results to determine these kind of systems mathematically. For standard systems that use traditional racks and single-load storage and retrieval machines, results have been presented since the 1970s. Nowadays, official guidelines from F.E.M exist.

Due to their popularity, the efficiency of automated storage and retrieval systems is continuously improved by various new applications and adaptions that are developed by practitioners in industry. In that way, industrial progress has overtaken academic research in some points. As a result, for some new applications, a fundamental mathematical description is not existing so far.

One possibility to increase the efficiency of automated storage and retrieval systems is to install a double deep rack structure and to use a storage and retrieval machine with two load handling devices. This allows enhanced space utilization and increased throughput potential. Figure 1.1 gives an example of this specific AS/RS design. Although such systems are applied in practice, we encounter the absence of feasible analytical formulations. Therefore, it is highly important to gain knowledge about this kind of AS/RS.

Figure 1.1: Example of an AS/RS with double deep storage and a dual load handling device

The objective of this thesis is to close this gap and to provide profound research in the field of AS/RSs which allow double deep storage and operate with dual capacity load handling devices. A general valid, **analytical travel time formula** for a random quadruple command cycle with double deep storage systems is developed. We study **routing and sequencing policies** (referred to as *strategies*) to improve throughput compared to a random execution of requests and mathematically derive related **models for quadruple command cycles** that consist of two storage and two retrieval operations. For selected strategies, we evaluate their potential for reduction of travel times with **extensive simulation studies**.

1.1 Problem Description

We differentiate two major research segments of our examination of double deep, dual load handling AS/RSs, resulting in two leading research questions.

First segment: Existing models and approaches of travel time determination for a quadruple command cycle in double deep storage systems have significant limitations. They do not allow for total randomized storage, all possible orders of execution and the different rearrangement options. For the first segment of research we formulate the leading research question.

> **Question 1: How can the mean travel time of a randomly executed quadruple command cycle be determined?**

Related questions are:

- What are essential assumptions for a randomly performed quadruple command cycle?
- How can a randomly performed quadruple command cycle be defined?

Second segment: Promising approaches of routing and sequencing strategies to improve throughput in AS/RSs are already known. While for traditional single deep AS/RSs analytical models exist, this is not the case for double deep environments. As available models are formulated with different assumptions and conditions, no consistent basis for a comparison of all possible strategies exists. The second segment of research is defined by the leading research question.

> **Question 2: How can sophisticated operating strategies be described and in which cases do they provide a travel time advantage?**

Related questions are:

- Which circumstances and conditions are relevant for the implementation of these strategies?
- How do they change in performance under consideration of sequencing of retrieval requests, demand structure of stock keeping units (SKU) or technical restrictions?
- How can they be analytically described? Can their travel times be determined mathematically?

1.2 Scope of this Thesis

This thesis is divided into six chapters that are structured in the following way:

In **Chapter 2** we provide basic knowledge on automated storage and retrieval systems. First we describe the structure and technical characteristics of AS/RSs. Important fundamentals with regard to organizational and qualitative descriptions are presented. Subsequently, the operation mode and common policies for storage, retrieval and sequencing are presented.

Chapter 3 focuses on the quantitative characterization of AS/RS. First, a formal description of storage systems is presented and mathematical basics are recapitulated. The derivation of fundamental travel time models, which are used to evaluate performance of AS/RS, are explained within a literature review. The second part of this chapter provides a review of existing research in relation to this thesis. In doing so, the research gap is emphazised as we present what is missing in existing research.

Building on this, **Chapter 4** covers the first research segment. We formulate a general analytical travel time model for the quadruple command cycle in double deep storage systems with dual load capacity load handling device. Using a parameter that describes the execution order of a quadruple command cycle, all possible executions order are included. The validity of the approach is shown by means of a simulation model. The chapter closes with a comparison to existing approaches.

Chapter 5 and 6 address the second research segment. Strategies are composed from existing approaches found in literature as well as from practical experience and a classification of strategies in a double deep storage environment with a dual capacity load handling device is developed. For selected strategy groups, analytical formulations are derived. In doing so, we provide basic components to describe many additional strategies in a mathematical way. To assess the strategies in consideration of real-world cases, a simulation model is used to compare strategies for various configurations and settings. Finally, we discuss results of our simulation studies as well as implications concerning implementation and application of the strategies in different environments.

Figure 1.2 shows the methodical categorization of the main parts and their coherence. While Chapter 4 follows a quantitative approach and Chapter 6 presents an empirical examination, chapter 5 combines both aspects. Composition and classification of strategies is a qualitative task, modeling their travel times based on the results of Chapter 4 is a quantitative task.

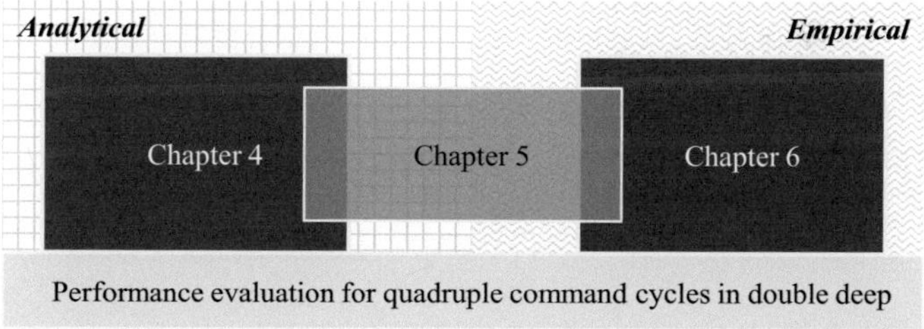

Figure 1.2: Methodical categorization of the main chapters

Finally, the results and accomplishments are summarized in **Chapter 7**.

2 Basics of Automated Storage and Retrieval Systems (AS/RSs)

We think too much
and feel too little.
-C. Chaplin

Warehouses are an inherent part of all flow of goods. They are required due to many different reasons such as ensuring supply of material, balancing different fluctuations of supply and demand, sorting and composing of orders, processing or venturing (Martin 2014, p. 336). At the same time, warehouses differ depending on the particular circumstances and requirements. We do not consider different motivations and functions of warehousing in greater details, while interested readers are recommend to study Martin (2014), Arnold, Isermann, Kuhn, Tempelmeier and Furmans (2008), Rushton (2017) and Bartholdi and Hackman (2016).

The focus of this chapter is to provide basic knowledge in the context of automated storage and retrieval system (AS/RSs). First, we introduce how different warehouse types can be categorized and how such systems can be classified. Afterwards, we present the technology of an AS/RS and explain how such systems are typically operated.

In literature, there are many ways of grouping warehouses, for examples by

- The function or the location of the warehouse within the supply chain,

- The type of the goods stored,
- The type of the unit load, e.g. pallets, bins or boxes,
- The warehouse design,
- The technology applied to handle goods.

The classification of warehouse design or warehouse type is not consistent in literature. A widely used way is to categorize warehouses into three groups: Floor storage, static rack- and dynamic rack storage (Arnold, Isermann, Kuhn, Tempelmeier and Furmans 2008, p. 648 ff.). Floor storage can be further divided into block or line storage, done by stacking pallets accordingly (Arnold and Furmans 2009, p. 191). Static rack storage refers to all types of systems where the storage unit remains at the same position. They are suitable for different types of unit loads. Common examples are shelving storage systems, adjustable pallet racking and high bay storage systems (Rushton 2017, p. 24 ff.). In dynamic rack systems, storage units are moved between put-away and picking, either by movement of the storage units itself or by movement of the rack. Examples of such types are flow racks, mobile racking systems or carousel systems. Apart from a complete manual operation, handling of goods can be realized in different ways using different technologies such as traditional fork lifts, other types of fork lift trucks (e.g. reach trucks and narrow-aisle trucks), storage and retrieval machines and special equipment such as robots and cranes (Arnold and Furmans 2009, p. 190). Figure 2.1 summarizes how automated AS/RS are defined in regard to warehouse design and handling technology.

AS/RSs are fully automated, computer controlled systems with fixed-path stacker cranes, also referred to as storage and retrieval machines (S/R machines), serving a static storage rack.

Note that another type of AS/RS, which gained a lot of attention in recent time, are systems using autonomous vehicle storage and retrieval systems (Roodbergen and Vis 2009). In these systems, horizontal and vertical transport of loads is separated from each other: The rail-guided vehicles move horizontally along aisles and cross-aisles. Lifting mechanisms, often located at the front side of the rack, are utilized for the transport in vertical direction. Another common name of these systems is shuttle based storage and retrieval systems (SBS/RS), where the term shuttle is applied to the vehicles. Epp (2017) provide an extensive literature review and comprehensive throughput evaluations of such systems.

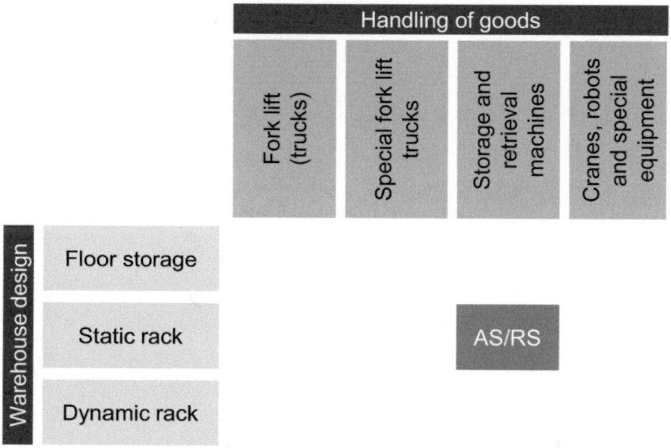

Figure 2.1: Characteristics of AS/RSs in terms of warehouse design and handling of goods.

2.1 Categorization and Functionality of AS/RS

Automated storage and retrieval systems are a preferred and commonly used warehouse type as they offer many advantages (Arnold, Isermann, Kuhn, Tempelmeier and Furmans 2008, p. 647) Due to a high degree of automation, labor costs are reduced whereas picking quality is increased compared to non-automated systems. Moreover, they provide a high space utilization which reduces the floor space required. Compared to other systems, AS/RSs yield high investment costs, e.g. for storage equipment and control systems, and less flexibility, which is why exact dimensioning is important. However, design and dimensioning depend on throughput requirements and thus **exact throughput determination** is crucial to make an adequate **design choice**.

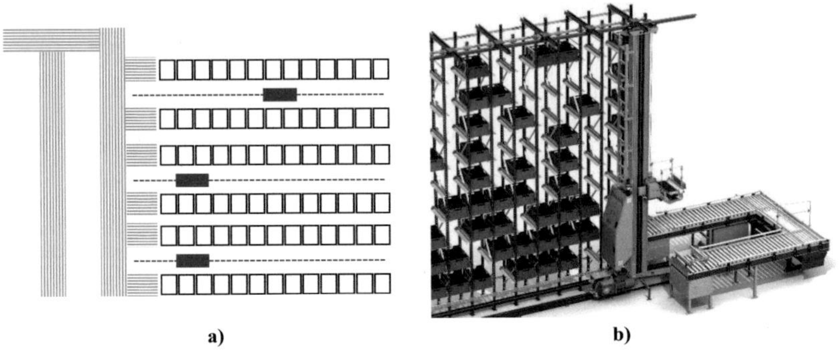

Figure 2.2: a) Top view of a schematic illustration for traditional AS/RS b) Example of an miniload AS/RS (Mecalux 2017a)

AS/RSs typically consist of racks, storage and retrieval machines with load handling devices, I/O points and a centralized control computer that is connected to the warehouse management system. S/R machines run up and down the aisles between the racks to pick-up and deliver loads. The racks consist of metal structures, forming the storage locations for the loads. The I/O points are location where loads are picked up before their storage and dropped of after being retrieved from the rack.

The traditional version of an AS/RS is a fully automated, unit-load, aisle-captive (one S/R machine per aisle) system with single-deep storage locations. Figure 2.2 a) shows a schematic illustration of the structure of a traditional AS/RS in a top view. 'Miniload' systems are the small versions of AS/RSs and basically consist of the same elements. Generally, they are lighter in structure, and achieve much higher values of velocity and acceleration (Arnold, Isermann, Kuhn, Tempelmeier and Furmans 2008, p. 666). Part b) of Figure 2.2 shows the side-view of a single aisle in a miniload system with plastic boxes as handling units. To give an overview of the various system options for AS/RSs, Roodbergen and Vis (2009) present a comprehensive classification of AS/RS. Figure 2.3 shows a modified classification derived from theirs.

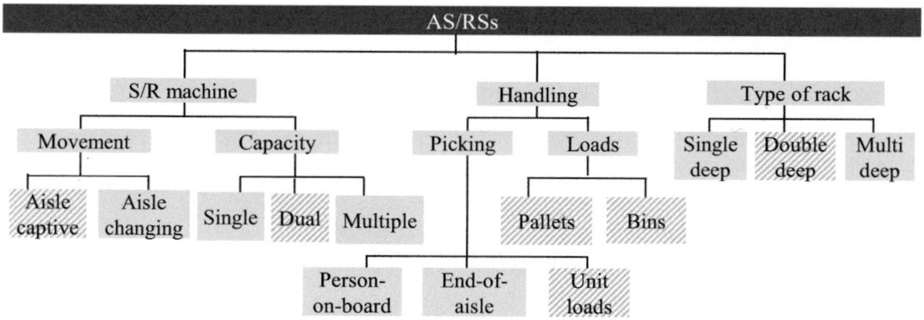

Figure 2.3: Possible classification of AS/RS variants adapted from Roodbergen and Vis (2009)

Based on Figure 2.3, the elements of AS/RSs are explained in the following. In doing so, we give an overview of the possible characteristics in each group and explain relevant details for the configuration applied in this work.

The models we formulate in this thesis apply to the blue shaded configurations: Fully automated unit load systems with stationary, double deep racks. The S/R machine is aisle-captive and has a dual capacity load handling device that is arranged in parallel and allow independent handling of the loads. Both, pallet and small part AS/RSs are relevant in this context.

2.1.1 Type of the rack

The rack construction is the core of the storage system with one or multiple aisles arranged in parallel. The design depends on individual requirements as there is no general implementation in terms of storage height, storage depth and number of aisles. AS/RS typically are operated in 'high bay storage racks' referring to heights from 12 meters (40 ft.) up to 50 meters (160 ft.). Rack lengths can range between 30 meters (99 ft.) and 150 meters (490 ft.) (ten Hompel and Schmidt 2010, p. 107). Those storage systems often serve as the building's supporting structure where the rack supports roof and side walls. Gudehus and Hofmann (1973) consider the ideal relation of height and length per pallet space, incorporating investment costs for the steel construction. The dimensions of the rack system influence the total number of storage positions. The determination of the needed number of storage positions depends on the inventory development over time for the different types of goods (Arnold and Furmans 2009, p. 176 f.).

However, design of the storage rack takes into account several more aspects, such as the direction of storage, i.e. is longitudinal or lateral to the rack, or the number of shelf uprights and the possible number of storage locations between them. The positioning of input and output locations that can be separated from each other, e.g. at different levels of height, is another design choice. For further explanation see ten Hompel and Schmidt (2010, p. 77), Martin (2014, p. 369) and Rushton (2017, p. 262).

For stationary racks operated by S/R machines, a distinction is made between single-, double-, and k- deep.

Single deep storage racks

Single deep storage systems represent the standard and the most common case of line storage with racks. All storage units are positioned next to each other within one plane where every unit is directly accessible from the aisle. Both, Figure 2.2 a) and Figure 2.2 b) show single deep storage.

Double deep storage racks

In double deep storage systems, two storage positions are located behind each other. Together, the position at the front, located next to the aisle, and the position behind form a storage lane. Figure 2.4 shows the top view of an aisle with double deep storage positions on both sides. Both positions of a storage lane are accessed from the aisle. The load handling device is able to reach both positions, through telescopic forks or similar mechanisms.

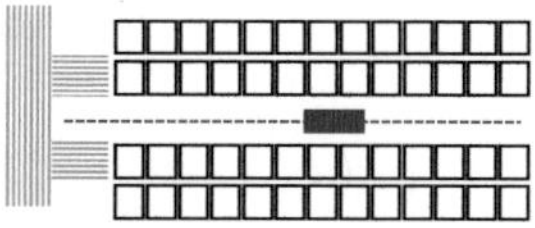

Figure 2.4: Top view of an aisle with double deep storage racks

Due to the reduced number of aisles, double deep storage allows denser storage and therefore can accommodate between 30% (Arnold, Isermann, Kuhn, Tempelmeier and Furmans 2008, p. 380) and 40% (Bartholdi and Hackman 2016, p. 54) more storage positions at the same floor space. However, direct access to the rear position is not always possible. If both storage positions of a storage lane are occupied, the rear position can only be accessed by rearranging the load at the front position. Consequently, the average access time is decreased because of such rearrangement operations.

Multiple deep storage racks

A k-deep storage system consists of k storage positions in one storage lane, also referred to as channel. As load handling devices with telescopic technology are limited in their range, different technologies are needed to serve all positions. Therefore, load handling for multiple deep storage, is different from those for single- or double deep racks in most cases. In the context of AS/RSs, this is typically achieved by shuttle based systems (Rushton 2017, p. 254). Multiple deep storage offers the advantage of high space utilization. However, especially if a channel can not be allocated to a single SKU type, a high number of rearrangement operations is provoked. Consequently, multiple deep storage is suitable for a small variety of different goods to efficiently use the capacity of channels.

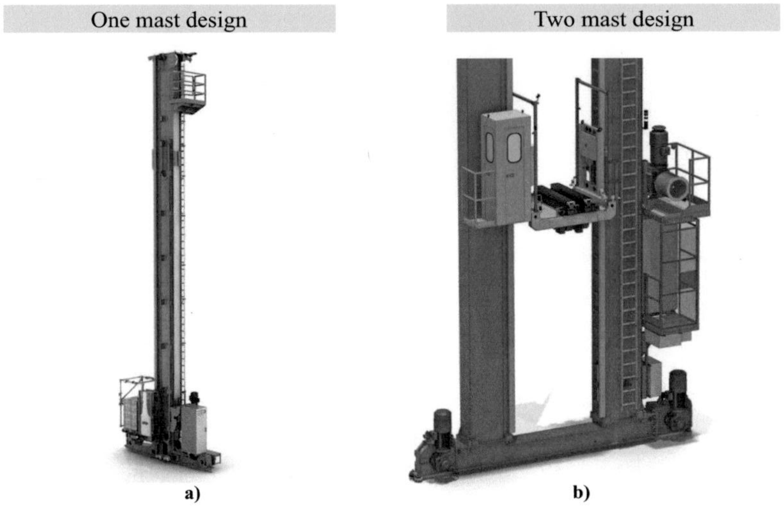

Figure 2.5: Example of one mast and two mast design AS/RSs Left: One mast design (SSI Schäfer Fritz Schäfer Gmbh 2017) Right: Two mast design (Mecalux 2017b)

2.1.2 S/R machine

Storage and retrieval machines are rail-guided on the floor of the aisles and at the top of the storage structure to maintain alignment. They consist of the following elements: Structural frame (rails and track), mast, lifting engine, load handling device, driving engine for horizontal movement, control system and in some cases they can be equipped with a cabin (Martin 2014, p. 389). Cabins can be used for emergency control or as manual picking systems.

The lifting engine works by a revolving mechanism such as drive belt, chain or cable. Horizontal movement is accomplished with friction wheel drive or drive belt, guided and supported by rollers (Arnold, Isermann, Kuhn, Tempelmeier and Furmans 2008, p. 664). Both engines, for vertical and horizontal movement, work independently of each other and can either be used separately or simultaneously. If both engines are powered on at the same time, simultaneous vertical and horizontal movement is possible.

The mast is usually made of aluminum or steel, nowadays also constructions made of fiber composites exist (Gebhardt Fördertechnik Gmbh 2017). A distinction is made between one and two mast systems, with the latter ones needed to reach heights of 45 meters (147 ft.) and more. The load handling devices of two mast systems are mounted between the mast, which allows them to bear greater loads, but also makes them heavier and more expensive. Figure 2.5 gives an example of both an S/R machine with a one-mast design in part a) and with a two-mast design in part b).

The relation between the number of aisles and the number of S/R machines is not necessarily one to one. If one S/R machine serves more than one aisle, it is able to change aisles by curved rails or with transfer cars that are located at the front side of the racks. We focus our study on aisle-captive systems that have one S/R machine in each aisle.

Capacity

Development of AS/RSs has constantly been evolved over the past decades. Nowadays, systems for dual and multiple load handling exist. In most cases, multiple load handling occurs along with two-mast systems. Figure 2.6 shows two possibilities of a dual load handling device. Figure 2.6 b) depicts a dual load handling device with two independent forks for pick-up and deposition. Figure 2.6 a) illustrates dual load handling with a double deep load handling device, which requires a double deep aisle (Arnold, Isermann, Kuhn, Tempelmeier and Furmans 2008, p. 666). Here, the units are positioned behind each other instead of side by side. Another option for dual load handling are load handling devices where the loads are positioned above each other. In the remainder of this work, we assume the load handling devices to be arranged in parallel and independent of each other as shown in Figure 2.6 b).

Miniload systems often provide multiple load handling devices of four or six units (Rushton 2017, p. 262).

Note that the load handling device is often referred to as shuttle in literature, especially if it is capable of loading more than one unit (e.g. 'dual-shuttle'). Nonetheless, storage and retrieval systems using shuttle vehicles, referring to tier captive or aisle captive satellites, are called shuttle based systems. For clarity, we use the term **load handling device** (LHD) instead of shuttle in the context of handling and movement of loads.

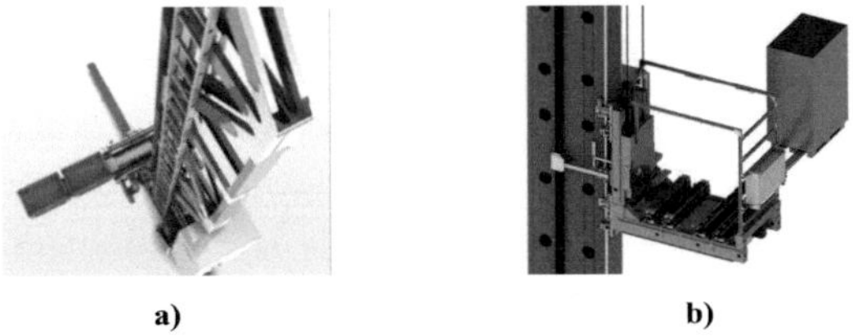

Figure 2.6: Two possibilities to realize dual load handling. a): With the loads behind one another in a double deep aisle (Gebhardt Fördertechnik Gmbh 2017) b): Loads next to each other in the direction of the aisle
(MIAS Group 2017)

2.1.3 Handling

Usually, AS/RSs handle goods in standardized containers. Two main types of loads can be distinguished: Pallets on the one hand and bins on the other hand. For the picking of loads, we can distinguish between picking of only full unit loads and parts of it. If partial unit loads are required, there are two common possibilities that integrate handling of partial amounts in the AS/RS.

Cabins that are mounted at the mast of the S/R machine represent one possibility to allow manual picking. In this set-up, referred to as person-on-board system, workers travel to the storage positions and pick-up single items from the loads, while the loads remain in the rack.

For the second option, loads are automatically retrieved and delivered to a picking workstation. There, workers take out single items before the load is stored back in the system. The workstations are often located at the front-side of the rack and referred to as end-of-aisle systems.

In the following, we consider the handling technology of unit loads. By handling technology we refer to the mechanisms used to retrieve the loads from and deposit the loads into the rack.

Handling of pallets is usually done with telescopic forks that extend under the pallet, lift them shortly by the lifting engine and then retract. For double deep storage, handling of pallets requires more space in vertical direction because of greater bending of the forks.

There are different technologies for the handling of bins. The possibilities are pushing/pulling mechanisms with fingers that move the load or grabbing the load with telescopic side clamps. Another option are fork-like telescopes that extract under the units, if possible, or use belt conveyors.(Arnold, Isermann, Kuhn, Tempelmeier and Furmans 2008, p. 666 f.)

2.2 Operating Policies of AS/RS

AS/RSs automatically perform storage and retrieval operations controlled by a computer system. Basically, loads are picked-up at the I/O point and moved to their storage location, whereas units to be retrieved are picked-up at their storage location in the rack and are moved to the I/O position where they are deposited. There are a number of control policies that can determine the possible actions performed by an AS/RS (Roodbergen and Vis 2009). This section explains the basic operating modes of an AS/RS and gives an overview of different policies. There is no standardized set of policies in the context of AS/RSs design and operation. In literature, policies are presented with different categorizations and levels of detail. Some of the most common control policies are routing policies, sequencing policies and storage assignment policies. Roodbergen and Vis (2009) also consider batching and dwell-point policies. Vasili, Tang and Vasili (2012) also incorporate load shuffling. Furthermore, Kraul (2010) and Atz (2016) distinguish between storage policy, retrieval policy, idle time policy and aisle changing policy. These groups are not mutual exclusive. Roodbergen and Vis (2009) provide an extensive summary of literature and a comprehensive description for AS/RSs' control policies.

We address the question of how the performance of an AS/RS (operating under different control policies) is determined based on travel times models in Chapter 3.

2.2.1 Storage Assignment

A storage assignment policy is a rule that determines how to choose storage positions within the rack. In literature, most attention is paid to random storage assignment and class-based storage assignment. Less common are dedicated storage assignments and full-turnover based storage assignments, whereas closest open location storage assignments occurs more rarely. At dedicated storage assignment, each SKU type is assigned to a fixed number of storage locations that are exclusively used for this type. Therefore, for each SKU there must be enough space available for the maximum number of units that may be stored at the same time. For closest open location storage assignment, the nearest empty location with respect to the I/O point is used. Full turnover based assignment is a special case of class based storage assignment which is discussed below.

Random Storage Policy

In random storage assignment, every load can be assigned to each available storage position within the rack. Consequently, each empty position has the same probability of being chosen as storage position and the units are evenly allocated throughout the rack. Due to pooling effects, i.e., the balancing of peak demands for different SKUs, less total storage space is needed compared to dedicated storage assignment. Moreover, random storage assignment requires low organizational effort and is a baseline for strategies that optimize throughput and cycle times.

Class Based Storage Policy

With classed based storage policies, all SKUs are analyzed and grouped according to their demand frequencies. The available storage space is divided into a certain number of classes, each belonging to one of the groups. The objective is to decrease mean travel times by locating those with the highest demand frequency closest to the I/O point. Within each class, random storage assignment is applied.

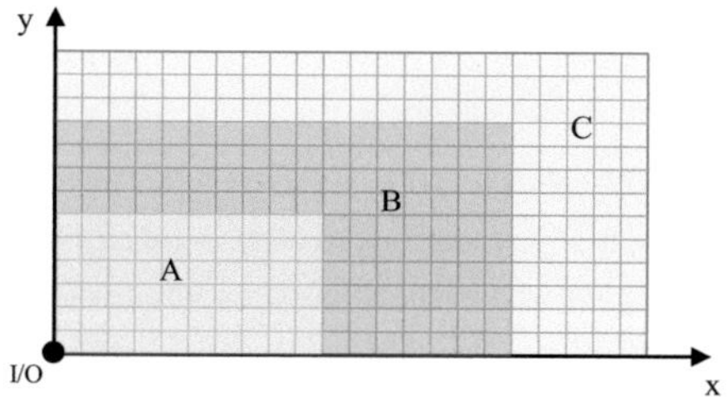

Figure 2.7: Example of class based storage policy with a rack divided into three different zones

The challenges are to determine the number of classes, to identify the number of products assigned to each class and to determine the location of each class within the storage rack. A common practice is to divide SKUs by means of an ABC analysis into three groups and use the result for the classification of the rack as shown in Figure 2.7 (Roodbergen and Vis 2009, p. 349). The aim of this approach is to limit the constitution effort for the class definition. Glass (2009, p. 47) emphasizes that many authors conclude that with such a small number of classes a majority of the potential for optimization is covered. In the extreme case of one class for each individual SKU, class based storage assignment turns into full turnover based storage assignment. In this case, the advantages of the random storage policy within the classes disappear, whereas organizational requirements increase (Kraul 2010, p. 48).

2.2.2 Routing and Sequencing

To provide a better understanding of routing and sequencing, we first explain the fundamental types of operation, before we go into more detail with routing and sequencing.

Type of operation

The Type of operation defines which command cycle is applied to operate an AS/RS.

Traditional AS/RSs that have a single capacity load handling device perform single or dual command cycles. In a **single command cycle** (SC), either a single storage or a single retrieval operation is performed. A storage cycle consists of picking up the load at the I/O point, traveling of the S/R machine to the storage position, placing the load into the rack and returning back to the I/O point. A retrieval cycle is performed similarly. Consequently, during a storage cycle the return travel is an empty run of the S/R machine, while for a retrieval cycle, the travel to the storage position is an empty run.

In a **dual command cycle** (DC), storage and retrieval are combined in one cycle as shown in Figure 2.8.

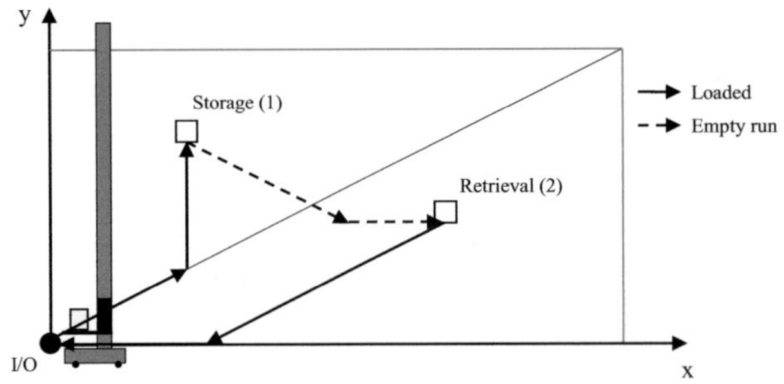

Figure 2.8: Example of a dual command cycle showing the movement of the S/R machine

In a first step, the load is picked-up at the I/O point, transported to its storage location and stored into the rack (1). Subsequently, the S/R machine travels to the retrieval location (2), the load is picked-up and returned to the I/O point. The travel distances within the rack, here between (1) and (2), are also called **travel-between** distances. Figure 2.8 shows a schematic side view of a storage rack and gives an example for the movement of the S/R machine while performing a dual command cycle. This way, empty travels as well as the time needed per operation can be reduced (Bartholdi and Hackman 2016, p. 202). While both storage and retrieval requests are available, it is always beneficial to perform dual command cycles to increase overall throughput. However, in cases when storage or retrieval are critical, performing only one type of a single command cycle may be appropriate, because the throughput of the individual type of operation (storage or retrieval) decreases with dual command cycles.

In a system with multiple load handling, command cycles of higher order are possible. With a dual capacity load handling device, two storage operations and two retrieval operations are possible in a single cycle, which is called a **quadruple command cycle** (QC). Similarly, for a load handling capacity of three units, **sextuple command cycles** become feasible by storing and retrieving three units each. Figure 2.9 illustrates a simplified example for the procedure of a quadruple and a sextuple command cycle. In both examples of Figure 2.9, all storage operations are performed first. However, the order can change, because retrieval operations are possible with at least one empty load handling device.

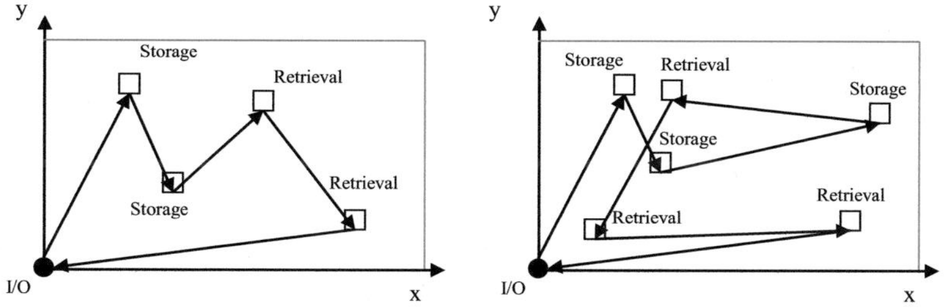

Figure 2.9: Simplified example of a quadruple and sextuple command cycle

Storage and retrieval selection

Different degrees of freedom are possible for the selection of storage and retrieval locations of an upcoming command cycle. In this thesis we assume the following:

Freedom of **storage selection** is determined by the storage assignment policy (see subsection 2.2.1), assuming more than one possible open location. The storage units remain in their order of arrival . Rearranging the order requires adequate space and conveyor technology, because they are represented by physical units (Kraul 2010, p. 43). Moreover, storage requests are not time-critical in general (Roodbergen and Vis 2009). As a consequence we do not consider the rearrangement of the order of storage requests.

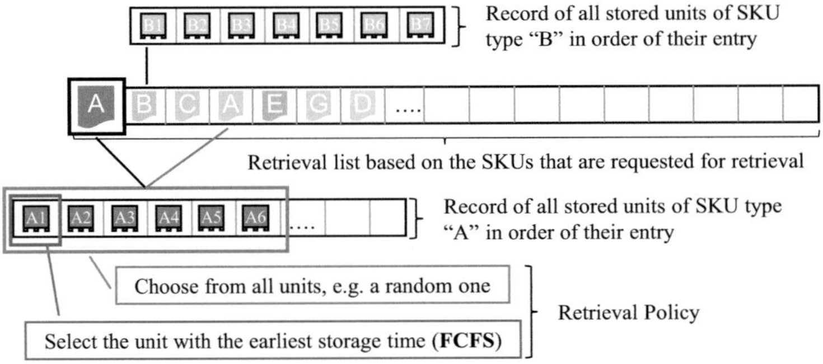

Figure 2.10: Illustration of the different aspects of retrieval selection

Freedom of **retrieval selection** is determined by two aspects that are illustrated in Figure 2.10.

1. The first aspect refers to the possibility of **sorting the list of retrieval requests**. This list represents the queue of upcoming retrieval jobs. Without this possibility, requests are processed in order of the entry into the list, which can be seen as a random execution.
2. The second aspect is referred to as **retrieval policy** (also called access policy), which describes the order of consumption within each SKU type (Atz 2016, p. 68f.). As every entry in the retrieval list represents

the demand for a specific SKU, this policies ensure that the selection of items is done in a desired way, e.g. 'First come first served' (FCFS), 'Last-In-Last-Out' or randomized selection (Kraul 2010, p. 42f.).

Understanding of routing and sequencing

Routing is the determination of the particular travel path in one command cycle. The possibilities for the path depend on the number of degrees of freedom for storage and retrieval selection, on the one hand, and on the number of stops in one cycle, on the other hand.

Sequencing is the examination of a series of command cycles as a tour to minimize the total time of all cycles (Roodbergen and Vis 2009). Usually, sequencing is done by cleverly selecting storage and retrieval locations. The greater the freedom for storage and retrieval selection is, the more sequencing options, i.e. possible combinations of storage and retrieval locations, exist. Dynamic sequencing is a special case in which the list constantly changes as new retrieval requests enter and thus ongoing re-sequencing is required.

To clarify these concepts, we use an example:

Example 2.1 *Consider a traditional AS/RS with a single load handling device that operates in a dual command cycle. Now, focus on the path determination of one particular cycle.*

Case 1: Let us assume the exact storage location of the unit to be stored is assigned beforehand, e.g. because of dedicated storage. Retrieval requests are executed in order of their arrival, which is why the retrieval location is defined as well. This means, there is no degree of freedom for both storage and retrieval selection. Consequently, the path for the dual command cycle with two stops is pre-determined.

Case 2: Storage locations can be chosen freely from all empty storage positions. Retrieval request are determined equally to case 1. Due to the degree of freedom in storage selection, there are many possibilities to determine the travel path of the cycle. Precisely, all empty positions serve as a potential first stop.

Case 3: Storage locations can be chosen freely from all empty storage positions. Re-sorting of retrieval requests is possible, which represents an additional degree of freedom for retrieval selection. Depending on the number of retrieval requests in the list, a number of potential paths result. With n empty positions and m retrieval requests, $n \cdot m$ potential paths for the next dual command cycle as well as sequencing options exist.

We can see that routing and sequencing are closely related and thus are often examined jointly. (see e.g. Van den Berg (1999) and Rouwenhorst et al. (2000)) In fact, sequencing describes the more complex routing problems.

The routing and sequencing of (several) command cycles formulates an optimization problem of finding a tour with the minimum total travel time for a given number of positions (Bozer, Schorn and Sharp 1990). This type of problem is known as the Traveling Salesman Problem, which is NP-complete. The time to solve the problem increases quickly with the problem size, i.e., the number of locations to visit in a single tour (Domschke 2007, p. 19ff.). Therefore, heuristics are developed to find efficient solutions causing reasonable effort.

Common heuristics

Han, McGinnis, Shieh and White (1987) were among the first to consider routing and sequencing heuristics for AS/RS. They propose the **Nearest Neighbor** heuristic which selects pairs with minimum travel-between distances from a list of n retrieval and s storage locations. Moreover, they propose the **Shortest Leg** heuristic that aims to find storage locations in the so called **No-cost** zone. Locations that meet such requirements lie in an area for which the S/R machine does not need extra travel time while traveling to the retrieval location. Both concepts represent common sequencing rules as many approaches for routing and sequencing of AS/RS are based on these ideas. Eynan and Rosenblatt (1993) propose a heuristic where Nearest Neighbor is combined with class based storage assignment.

Due to the increased number of stops within a single cycle, higher order command cycles provide more possibilities of routing. A possible approach for multiple load handling is the **Flip Flop** heuristic. Under this policy, a location that became free trough retrieval can be subsequently used for storage (Sarker, Sabapathy, Lal and Han (1991), Keserla and Peters (1994)). Another one is the idea of **Multiple storage** which is to simultaneously store the loads next to each other, either in two neighboring storage positions or behind each other in one storage lane (Seemüller 2006). Relevant literature in this context is discussed in subsection 3.3.1.

3 Performance of Automated Storage and Retrieval Systems

Education is not preparation for life; education is life itself.
-J. Dewey

The objective of this chapter is to provide basic knowledge about performance evaluation of AS/RSs. The most relevant approach is the determination of mean travel times by travel time models that express the average time needed to perform a command cycle. Travel time models present a mathematical representation of the expected value of all possible cycles, given a particular storage rack configuration. Besides, there are additional ways of performance determination: For complex problems, no closed form expressions can be given. In this case, mathematical functions are numerically evaluated by a computer program or approximations may be used to obtain results for the original problem. Another option is to evaluate performance based on computer simulations that allow to perform a high number of randomly generated iterations. Thus, it is also possible to conduct experiments that are not achievable with real systems because of physical or economical restrictions. VDI Richtlinie 4480, Blatt 4 (1998) recommends to apply simulation studies for a consideration of more individual influences, e.g. of articles or order structures. Another method is to perform only one cycle that generates the mean travel time. Therefore 'representative positions' are used.

Next, we provide relevant mathematical basics underlying both travel time determination and the approach we follow on this thesis, as well as the fundamental modeling assumptions. In the second part of this chapter, important analytical travel time models as well as a broader range of studies in the context of performance evaluation of AS/RSs are discussed. Thirdly, we study relevant literature for our research.

3.1 Basics of Travel Time Determination

The performance of automated storage and retrieval systems is usually measured based on mathematical **travel time** or **cycle time** models. They are used to calculate the time of an average (single-, dual- or other) command cycle together with rearrangements for multi-deep storage systems.

The total cycle time consists of path-depending travel times and dwell times (Gudehus 1972d). **Path-depending** travel times refer to the movement between the storage and retrieval locations to be approached during the cycle. **Dwell times** are independent of the path-depending travel times and occur in every cycle. In accordance with Lippolt (2003) we define the following elements of dwell times:

- **Load handling time** (t_{LHD}): This is the time of one access cycle of the load handling device, consisting of extension and retraction of the LHD, e.g. to pick-up a unit from the rack or deposit a unit in the rack.
- **Dead time** (t_{dead}): Dead times account for the time needed for reaction, control and operation of sensors within a cycle. During dead times, no movement takes place. We assume two t_{dead} per access cycle of the load handling devices.
- **Mast damping period** (t_{mast}): Due to acceleration and deceleration, the mast of the S/R machine oscillates with every movement. Before the load handling starts, the oscillation needs to calm down. The related waiting time is referred to as the mast damping time and is depending on the mass of the mast. In general, this time occurs after every travel movement.

Moreover, travel times depend on the speed and both acceleration and deceleration of the S/R machine (Lippolt 2003, p. 46):

3.1.1 Mathematical Basics

First, we present the basics of order statistics, which are applied in many approaches of travel time modeling. As our central approach bases on stochastic processes, we subsequently explain the important concepts in this regard.

Order statistics

Order statistics are used to denote the ith smallest or i^{th} largest instant within a sample. Suppose X is a continuous random variable with a cumulative distribution function (cfd) $F_X(x)$ and a probability density function (pdf) $f_X(x)$ and the independent, identically distributed random samples $X_1, X_2, \cdots X_n$. The realization of these variables $x_{(i)}, i = 1, \cdots, n$ can be ordered such that $x_{(1)} \leq x_{(2)} \leq \cdots \leq x_{(n)}$. The order statistics $X_{(1)}, X_{(2)}, \ldots, X_{(n)}$ are random variables of the ordered values. The ith entry of the ordered sample is referred to as the ith order statistics. Note that $x_{(i)}$ is the realization that of the ith order statistics $X_{(i)}$ but, in general, not the realization of random variable X_i.

For the cumulative distribution function of the first order statistics we know:

$$F_{X_{(1)}}(x) = P(X_{(1)} \leq x) = 1 - P(X_{(1)} > x). \tag{3.1}$$

For the smallest value of the sample, $X_{(1)} > x$ holds if and only if $X_{(i)} > x$ for all $i = 1, 2, \cdots, n$. As the individual $X_{(i)}$ are stochastically independent, the smallest order statistics is (Devore and Berk 2012, p. 272 f.):

$$\begin{aligned} F_{X_{(1)}(x)} &= 1 - P(X_{(1)} > x) \\ &= 1 - P(X_1 > x)P(X_2 > x) \cdots P(X_n > x) \\ &= 1 - (1 - F(x))^n. \end{aligned} \tag{3.2}$$

Devore and Berk (2012, p. 276 f.) present a clear derivation of the ith order statistics and determine the pdf of the ith order statistics by:

$$\begin{aligned} f_{X_{(i)}(x)} &= nf(x)\binom{n-1}{i-1}F(x)^{(i-1)}(1-F(x))^{(n-i)} \\ &= \frac{n!}{(i-1)!(n-i)!}f(x)F(x)^{(i-1)}(1-F(x))^{(n-i)}, \\ & x \in \mathbb{R}, 1 \leq i \leq n. \end{aligned} \tag{3.3}$$

The cdf of the ith order statistic is (Zörnig 2016, p. 178):

$$F_{X_{(i)}(x)} = \sum_{k=i}^{n}\binom{n}{k}F(x)^k(1-F(x))^{(n-k)}, x \in \mathbb{R}, 1 \leq i \leq n. \tag{3.4}$$

The range of the sample, R, is the difference between the largest and the smallest value, i.e., $R = X_{(n)} - X_{(1)}$. According to Guttman, Wilks and Hunter (1982) the cdf and pdf, $H(r)$ and $h(r)$, respectively, of the range are defined as:

$$H(r) = n \int_{v=-\infty}^{\infty} f(x)[F(x+r)-F(x)]^{n-1}dx \tag{3.5}$$

$$h(r) = n(n-1)\int_{v=-\infty}^{\infty} [F(x+r)-F(x)]^{n-2}f(x)f(x+r)dx \tag{3.6}$$

Stochastic processes

Stochastic processes are used to describe real procedures or system behavior over time. Usually, all possible states and the transition from one state to another are analyzed. A stochastic process is a family of random variables $\{X_t, t \in T\}$ with the index set T. T usually is a set of points in time when the system is observed. For $T = \mathbb{N}_0$ the stochastic process is said to be discrete-time process and for $T = [0,\infty)$ a continuous-time process. In the context of this work, discrete-time processes are relevant which is why we refrain from going into detail for the second group. The possible states the random variables X_t can take are represented by the state space S. This means, a stochastic process is a sequence of random variables $X_0, X_1, \cdots$ that take values from S and are observed at the points in time T. The difference between a stochastic process and a random sample of X is that the sample values $X_1, \cdots, X_n$ are independent of each other whereas the random variables of the stochastic process are not.

A common example is queuing of customers waiting to be served at a counter. In this example, a random variable (X_t) is used to describe the state of the queuing system, which is the number of customers waiting in the queue. All possible numbers of waiting customers that can be observed is $S = \mathbb{N}_0$. Now consider a point in time t, where three waiting customers are observed ($X_t = 3$). The next point in time, $t+1$ (after a state transition), depends on the previous state: If one customer arrives without completing any of the already waiting customers in service, the new number of waiting customers is four ($X_t = 4$). On the other hand, if no customer has arrived and the ongoing service was completed, the number of customers waiting is reduced to $X_t = 2$. Obviously $t+1$ is dependent on t.

If the state of a stochastic process, t, is only dependent on state $t-1$ and not of previous ones, this is called **Markov Property**. A process with the Markov Property is also called memoryless.

Discrete Time Markov Chains

A stochastic process $\{X_t, t \in T\}$ taking values in a countable state space S is called Markov chain, if for a all points in time $t \in T$ and all states $i_0, \cdots, i_{t-1}, i_t, i_{t+1} \in S$ the following is true (Waldmann and Stocker 2013, p. 11):

$$\begin{aligned} P(X_{t+1} = i_{t+1} | X_0 = i_0, \cdots, X_{t-1} = i_{t-1}, X_t = i_t) \\ = P(X_{t+1} = i_{t+1} | X_t = i_t) \end{aligned} \tag{3.7}$$

This represents the **Markov Property** and is expressed through the transition probabilities:
The conditional probability $P(X_{t+1} = i_{t+1}|X_t = i_t)$ is called transition probability of the process and represents the probability that for a given state i_t the following state i_{t+1} is realized. This means that the transition probability $P(X_{t+1} = i_{t+1}|X_t = i_t)$ from state i_t into state i_{t+1} is only dependent of i_t, but of no other state prior to i_t .

For transition probabilities independent of the time of the transition t, the Markov chain is said to be **homogeneous.** The evolution of a homogeneous Markov chain can be described by

1. **Initial probability** $\pi_i(0) := P(X_0 = i), i \in S,$
2. **Transition probabilities** $p_{ij} := P(X_1 = j|X_0 = i), i, j \in S,$
3. **Transition matrix** $P = (p_{ij}),\ p_{ij} \geq 0\ \forall i, j \in S$
   $\wedge \sum\limits_{j \in S} p_{ij} = 1,\ \forall i \in S.$

The transition matrix is a stochastic matrix that describes the transitions of a Markov chain. As the entries of the square matrix represent transition probabilities, all entries are greater or equal zero and and all rows sum up to one.

$p_{ij}^{(n)} = P(X_n = j|X_0 = i) = \sum\limits_{i_1 \in S} \dots \sum\limits_{i_{n-1} \in S} p_{i,i_1} \dots p_{i_{n-1},j}$ is defined as n-step transition probability which denotes the probability of going from state i to state j in n transitions. $\pi_j(t)$ is the marginal distribution over states at time n, the probability distribution of the random variable $\{X_t, t \in T\}$ is described as:

$$\pi_j(t) = \sum_{i \in S} P(X_0 = i) P(X_t = j|X_0 = i) = \sum_{i \in S} \pi_i(0) p_{ij}^{(t)}. \tag{3.8}$$

$p_{ij}^{(t)}$ is obtained by adding up the probabilities of all sequences of states $p_{i,i_1} \dots p_{i_{t-1},j}$ $(i, i_1 \dots i_{t-1}, j \in S)$, beginning in i and ending up in j after t steps. As $p_{ij}^{(t)}$ is a conditional probability, it is multiplied by the initial probability of state i in equation 3.8. Summarizing all initial states $i \in S$, according to the law of total probability the unconditional probability $\pi_j(t)$

can be calculated (Waldmann and Stocker 2013, p. 15f). We can interpret the state probabilities $\pi_j(t)$ for all $j \in S$ as a row vector $\pi(t)$ giving the state distribution at time t.

It is possible to analyze the evolution of Markov chains for $t \to \infty$ to derive a stationary distribution. For this purpose, we want to discuss some properties of Markov chains:

A state i is said to **communicate** with state j, if they are accessible from each other. A Markov Chain is said to be **irreducible**, if all states communicate, i.e., for every state i there is a positive probability of going into state j. If an irreducible Markov chain has a finite state space, it has a unique stationary distribution (Waldmann and Stocker 2013, p. 36). A state i of a Markov chain is **recurrent**, if it has a finite return time, so with a probability of one the chain returns to state i after a finite number of transitions. A state i is called **aperiodic**, if the transition to the same state has a non-zero probability, which is $p_{ii} > 0, i \in S$. An irreducible Markov chain is aperiodic if it has at least one aperiodic state (Waldmann and Stocker 2013, p. 41).

Let $\{X_t, t \in T\}$ be an irreducible, aperiodic Markov chain with a stationary distribution π, then $\pi(t)$ converges to the stationary distribution for $t \to \infty$. The stationary distribution is independent of the initial distribution of the process. This means the stationary distribution is reached regardless of the starting point. A distribution is said to be stationary, if

$$\pi_j = \sum_{i \in I} \pi_i p_{ij}^{(t)}, \; j \in S, \; \forall t \in \mathbb{N}. \tag{3.9}$$

This could also be expressed as the convergence of the transition matrix in the following way:

$$\lim_{t \to \infty} p_{ij}^{(t)} = \pi_j > 0 \;\text{ for all } j \in S \tag{3.10}$$

Let $u_i \in [0,1], i \in S$ be a probability distribution of X_t. π is given as the solution of the following system of linear equations

$$u_j = \sum_{i \in I} u_i p_{ij},\ j \in S \tag{3.11}$$

$$u_i \geq 0,\ i \in S \tag{3.12}$$

$$\sum_{i \in S} u_i = 1 \tag{3.13}$$

π_j is the probability that the system is in state j for $t \to \infty$ and, as can be seen from equation 3.10, is independent of the initial state. After a sufficient period of time, π_j can also be interpreted as the mean proportion of time the system is in state j.

3.1.2 Modeling of the Storage Rack and the Movement of the S/R machine

In order to analytically determine travel times in an AS/RS, not the whole storage system is considered. For reasons of simplicity, a single pick face of a storage system operated by an automated S/R machine is considered. As an AS/RS typically is constituted of many subsystems, such as aisles or single pick faces, total throughput can be derived by the consideration of one subsystem, as long as they are operated in the same way.

Behavior of the S/R machine

The S/R machine moves between the I/O point and its maximum lifting height and travel distance. The maximum travel speed the system can reach in horizontal direction is v_x. The lifting mechanism is able to reach a speed of v_y in vertical direction. Usually, both engines operate at the same time, resulting in a simultaneous movement in horizontal and vertical direction of the load handling device. To achieve a one-dimensional movement, the engines need to be powered on individually.

Figure 3.1 depicts the speed-time behavior for both drive and lifting engine. The maximum possible speed is v_{max}.

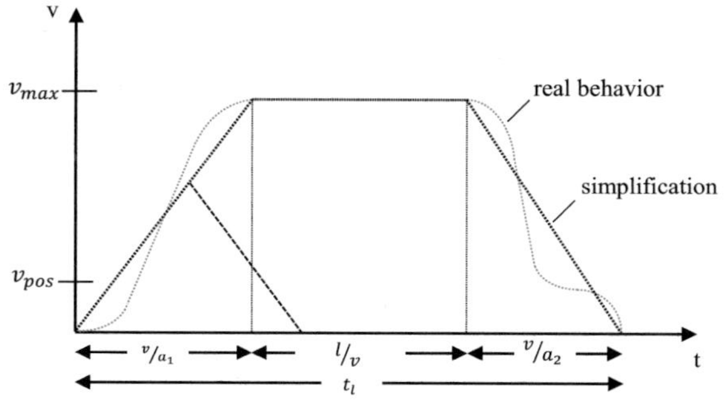

Figure 3.1: Speed-time graph showing the behavior of the S/R machine (Arnold and Furmans 2009, p. 204)

In the following, for the explanation of Figure 3.1 assume $v = v_{max} = v_x$ or $v = v_{max} = v_y$, respectively. The dotted line shows the real behavior, while the solid line represents the simplification where acceleration and deceleration are linearized and the positioning time at the end of the movement is ignored. Because of its shape, the simplified behavior is referred to as **trapezoid profile**. For short travel distances, where the maximum possible speed is not reached, the deceleration phase starts immediately after the acceleration phase leading to a **triangular profile** and a speed below the the maximum possible speed. The black, dashed line in Figure 3.1 represents this case. In case the values for acceleration, a_1, and deceleration, a_2, are different, they are averaged for simplification (Gudehus 1972d):

$$a = \frac{2|a_1 a_2|}{a_1 + |a_2|} \tag{3.14}$$

This results in acceleration and deceleration phases of equal length. The time that is needed to reach the maximum speed is given by

$$t_{acc.} = \frac{v}{a} \tag{3.15}$$

Consequently, the time needed to speed up to the maximum speed and subsequently slow down to a full stop is $2 \cdot t_{acc.} = 2\frac{v}{a}$. Let t_l be the travel time of the S/R machine between two locations. The distance, l, that is covered during t_l can be calculated from the simplified line by integrating over time:

$$l = \int_0^{t_l} v(t)dt = \begin{cases} \frac{a}{4} t_l^2 & \text{for } t_l < 2\frac{v}{a} \\ v t_l - \frac{v^2}{a} & \text{for } t_l \geq 2\frac{v}{a} \end{cases} \tag{3.16}$$

By transforming equations 3.16, the travel time t_l is given by:

$$t_l = \begin{cases} 2\sqrt{\frac{l}{a}} & \text{for } l < \frac{v^2}{a} \\ \frac{l}{v} + \frac{v}{a} & \text{for } l \geq \frac{v^2}{a} \end{cases} \tag{3.17}$$

For travel time determination, usually all distances are calculated for $l \geq \frac{v^2}{a}$ in equation 3.17. Thus, the triangle velocity profile is ignored. As a result, the real time is overestimated by $\frac{v}{a}$ at a maximum, but the overall error is negligible (Arnold and Furmans 2009, p. 205).

Rack model and movement within the pick face

The rack is represented as a **discrete** or **continuous** model. In the discrete model there is a given number of storage positions, each having a defined position. For the continuous model, an unlimited number of infinitesimal small storage positions is assumed. The latter approach offers the advantage of an analytical solution by integration instead of numerically incorporating all storage positions of the rack. The continuous modeling approach is the more common method for travel time determination because of its facilitated way of calculation and is also applied by Schaab (1969), Bozer and White (1984) and Gudehus (1972d).

Figure 3.2 shows a characteristic storage rack model with the I/O point in the bottom left corner at (0,0). The position of the farthest column or the maximum travel distance of the S/R machine is L and the position of the top row or the maximum lifting height of the S/R machine is H.

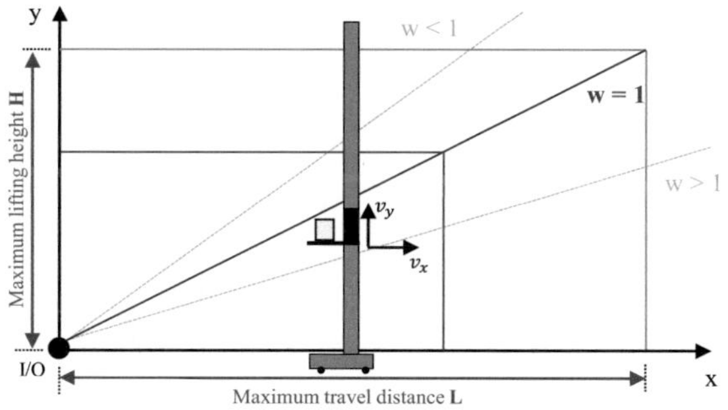

Figure 3.2: Rack model and S/R machine with a synchronous movement line for $w = 1$ (bold line) and example isochrone.

For the synchronous operation of both engines, the load handling device (one distinct position at the LHD) moves along a straight line, that is described by Gudehus (1972d):

$$y = \frac{v_y}{v_x} \cdot x \tag{3.18}$$

This so called **synchronous movement line** depends on the relation of the speed in horizontal and vertical direction. In combination with the dimensions of the rack, H and L, this line influences the travel times obtained in the rack. This effect is summarized in the shape parameter w, which is defined as follows:

$$w = \frac{v_x}{v_y} \cdot \frac{H}{L} \tag{3.19}$$

w describes the slope of the synchronous movement line. For $w = 1$, the S/R machine reaches the farthest position in the vertical (H) and the horizontal (L) direction at the same time. For a shape factor $w < 1$, the maximum height is reached before the maximum length is reached. To approach the top right corner of the rack, the engine for horizontal movement is activated longer than the lifting drive. Similarly, for $w > 1$, the farthest location in x-direction is reached before the highest location and the lifting time is determinant for the total travel time. As the smallest mean travel time is obtained for $w = 1$, the dimensioning of the system should be guided by that relation (Gudehus 1972b). Moreover, equation 3.18 bisects the storage positions of the rack: The storage positions above and below the diagonal line. For the first group, the travel time from the I/O point is only determined by the lifting engine whereas for those below it, the horizontal drive is crucial for travel time determination. For each location $P = (x, y)$ in the storage rack, the time to reach P is given by the maximum of the two one-dimensional movements, t_x and t_y respectively. Both are calculated according to 3.17. The path-depending travel time needed to reach P is:

$$t_l = max(t_x; t_y) \tag{3.20}$$

Let a_x and a_y be the S/R machine's acceleration and deceleration in horizontal and vertical direction, respectively. For the two-dimensional travel of the S/R machine, Gudehus (1972d) determines the impact of acceleration and deceleration, allowing for different positions above or beyond the synchronous travel line.

$$t_a = \begin{cases} (1 - \frac{w}{2}) \cdot \frac{v_x}{a_x} + \frac{w}{2} \cdot \frac{v_y}{a_y} & \text{for } w \le 1 \\ \frac{1}{2w} \cdot \frac{v_x}{a_x} + (1 - \frac{1}{2w}) \cdot \frac{v_y}{a_y} & \text{for } w > 1 \end{cases} \tag{3.21}$$

This yields $\frac{1}{2}(\frac{v_x}{a_x} + \frac{v_y}{a_y})$ in the mostly used case of $w = 1$.The combination of positions with either the same x-location below the synchronous movement line or the same y-location above it and $(x, y) \in w$, form an isochrone, i.e., the travel time, beginning in (0,0), is identical. One example of such an isochrone is shown in Figure 3.2.

Transformation of coordinates

Besides mapping the rack by Cartesian coordinates using the maximum length L and height H, it is also possible to indicate positions according to the travel time. The idea is, to scale the rack in relation to the maximum possible travel time instead of using the real measurements of the rack. In the statistical approach by Bozer and White (1984), the coordinates are transformed into time-scaled coordinates. To interrelate the dimensions of the rack and the kinematic characteristics of the S/R machine, the maximum travel time in every direction is calculated:

$$t_{x,max} = \frac{L}{v_x} \tag{3.22}$$

$$t_{y,max} = \frac{H}{v_y} \tag{3.23}$$

with $t_{x,max}$ representing the travel time to reach the farthest position in x-direction and $t_{y,max}$ representing the maximum lifting time to reach H. T is the normalization factor for the transformation of the rack

$$T = max(t_{x,max}; t_{y,max}) \tag{3.24}$$

and denotes the maximal travel time obtained for the given system.

The normalized shape factor of the rack, b, is defined as follows:

$$b = min\left(\frac{t_{x,max}}{T}; \frac{t_{y,max}}{T}\right) \tag{3.25}$$

As a consequence of 3.24 and 3.25, one has

$$0 \leq b \leq 1. \tag{3.26}$$

Figure 3.3 shows the scaled rack model, where the gray font describes the situation before the transformation. The rack is transformed from the distance measured rectangle with the size $(L \times H)$ into a dimensionless rectangle with the size $(1 \times b)$. Without loss of generality, let $t_{x,max} > t_{y,max}$ and therefore $T = t_{x,max}$ as well as $b = \frac{t_{y,max}}{T}$. Next, the dimension of the rack with the greater maximum travel time is scaled to 1, this means $t_{x,max} = 1$.

The coordinate of the other dimension is b. Consequently, the scaled rack has a size of $(1 \times b)$. The vertical dimension is by b smaller than the horizontal dimension, which is why b defines the shape of the transformed, normalized rack.

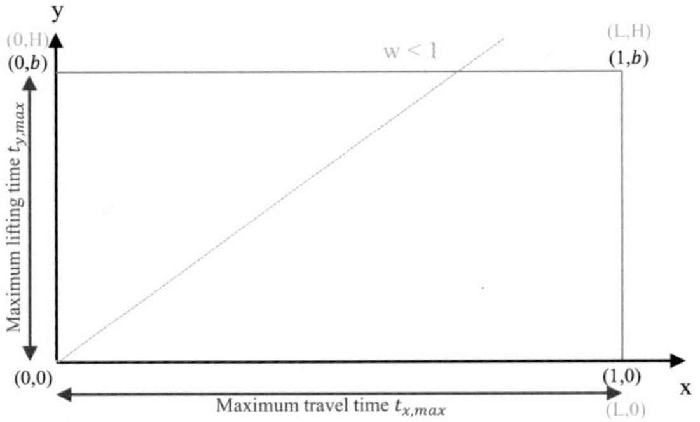

Figure 3.3: Scaled, dimensionless rack with time coordinates

b has a similar meaning to w, which follows from equation 3.19:

$$w = \frac{v_x}{v_y} \cdot \frac{H}{L} = \frac{t_{y,max}}{t_{x,max}} = \frac{t_{y,max}}{T} = b$$

For the given example, in the case of $t_{x,max} > t_{y,max}$, b equals w. For $t_{x,max} < t_{y,max}$, $b = \frac{1}{w}$ is valid. As w can be greater than 1, whereas b is defined according to 3.26, the relation between both is:

$$b = \begin{cases} \frac{t_{y,max}}{t_{x,max}} = w & \text{for } w \leq 1 \\ \frac{t_{y,max}}{t_{x,max}} = \frac{1}{w} & \text{for } w > 1 \end{cases} \tag{3.27}$$

The configuration $b = w = 1$ is called **square in time**, with the scaled and transformed rack being square-shaped (Bozer and White 1984).

Remember that in the non-scaled rack model, the travel time from $(0,0)$ to each location $P = (x, y)$, t_l, is determined according to equation 3.20.

For each location $P' = (x, y)$ in the transformed rack model, the normalized travel time from $(0,0)$ to P', t_n, is determined by:

$$t_n = max\{x; y\} \tag{3.28}$$

As the result is normalized by the maximum travel time possible in the rack, the scaling factor T is needed to derive real travel times.

Besides, this approach does not take acceleration and deceleration into account. The results must therefore be adjusted by the components of Gudehus (1972d) from equation 3.21 to incorporate those phases.

3.2 Travel Time Determination for AS/RSs

The first examination of travel time models for storage and retrieval systems goes back to Zschau (1963) and Schaab (1969) who determine mean single and dual command cycles. For both single command cycle and travel between distance, they define an integral formulation based on the infinitesimal consideration. Speed as well as acceleration/deceleration of the S/R machine are taken into account.

Graves, Hausman and Schwarz (1976, 1977, 1978) present expressions to determine the single and dual command cycle assuming the rack to be square in time ($b = 1$). They compare a random storage policy to class-based and full-turnover based storage policies. Moreover, they consider a policy similar to the nearest neighbor idea for the selection of retrieval jobs. They were among the first to show the potential of those policies and influenced many authors in the upcoming years to further research.

However, the commonly referenced, fundamental travel time models are derived by Gudehus (1972d) and most important Bozer and White (1984).

3.2.1 Fundamental travel time models

Gudehus (1972d) and Bozer and White (1984) independently from each other develop travel time models for single and dual command cycles that incorporate storage racks that are not square in time, also allowing for alternative I/O points. They follow two different approaches in the modeling of the storage rack, but their results can be transferred into each other and are also in line with the achievements of Graves, Hausman and Schwarz (1977).

In the following we present the derivation of the travel time based on the approach of Bozer and White (1984) and subsequently compare them to results of Gudehus (1972d) to show their consistency.

Bozer and White (1984) derive the mean travel times for single and dual command cycles based on a statistical approach. They use the transformed model for the storage rack with normalized coordinates representing travel times. Moreover, they assume a randomized storage policy, i.e., any open position in the rack is equally likely to be selected for storage and any occupied position is equally likely to be selected for retrieval. They require the S/R machine to operate at full utilization, meaning no waiting times of the S/R machine occur. They neither incorporate any dwell times nor acceleration/deceleration.

Derivation of the Single Command Cycle

To determine a single command cycle, the one-way travel from the I/O point to a randomly chosen position in the rack, $P = (x, y)$, is considered, referred to as $E(SW_1)$. $T = t_{x,max}$ and $b = \frac{t_{y,max}}{T}$ are assumed, meaning the transferred rack dimensions are $(1 \times b)$ as shown in Figure 3.4 (The opposite case, $T = t_{y,max}$ and $b = \frac{t_{x,max}}{T}$, leads to the same result). Hence, for x and y applies

$$0 \leq x \leq 1 \tag{3.29}$$

$$0 \leq y \leq b \tag{3.30}$$

and according to equation 3.28 the time needed to travel to that point is $t_n = max\{x; y\}$.

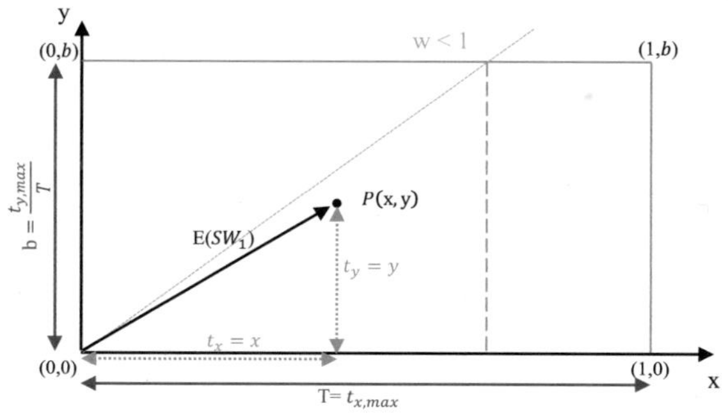

Figure 3.4: Travel time determination between I/O and a random point $P = (x, y)$ according to Bozer and White.

Bozer and White (1984) formulate the condition that the travel time is less or equal to $\zeta \in [0,1]$. Because x and y are independent of each other, this can be expressed as follows:

$$G(\zeta) = P(t_n \leq \zeta) = P(x \leq \zeta)P(y \leq \zeta) \tag{3.31}$$

Due to the randomized storage policy, the positions of x and y are uniformly distributed across the pick face which leads to:

$$P(x \leq \zeta) = \zeta \tag{3.32}$$

$$P(y \leq \zeta) = \begin{cases} \frac{\zeta}{b}, & \text{for } 0 \leq \zeta \leq b \\ 1, & \text{for } b \leq \zeta \leq 1 \end{cases} \tag{3.33}$$

As b limits the scaled rack in y-dimension, the probability has to be split into the two different cases. The upper refers to the situation, when ζ is less than b. Accordingly to the uniform distribution, this probability is equally spread over the range of b. The lower part of equation 3.33, for ζ being greater than b, has a probability of 1. As the y-dimension is limited with b, y is always smaller than ζ.

With these probabilities, according to equation 3.31 $G(\zeta)$ is:

$$G(\zeta) = \begin{cases} \frac{\zeta^2}{b} & \text{for } 0 \leq \zeta \leq b \\ \zeta & \text{for } b < \zeta \leq 1 \end{cases} \tag{3.34}$$

When differentiating this distribution function, the pdf is

$$g(\zeta) = \begin{cases} \frac{2\zeta}{b} & \text{for } 0 \leq \zeta \leq b \\ 1 & \text{for } b < \zeta \leq 1. \end{cases} \tag{3.35}$$

The expected one-way travel time is:

$$E(SW_1) = \int_0^1 \zeta g(\zeta) d\zeta = \frac{1}{6}b^2 + \frac{1}{2} \tag{3.36}$$

The expected single command cycle time is the time for a return travel to a random position, so therefore:

$$E(SC)^N = 2E(SW_1) = \frac{1}{3}b^2 + 1 \tag{3.37}$$

For racks that are square in time, having $b = 1$, $E(SC)^N$ is 4/3. The result is normalized and dimensionless, indicated by the superscript N. To obtain time-scaled results, one needs to multiply with the scaling factor T (see equation 3.24):

$$E(SC) = (1 + \frac{b^2}{3})\ T \tag{3.38}$$

Under the assumption of a random storage policy and the I/O point being in the bottom left corner, Gudehus (1972d) present the following results for expected mean single command cycle time with w and t_a from equations 3.19 and 3.21.

$$E(SC) = \begin{cases} t_0 + 2t_a + \frac{L}{v_x}[1 + \frac{w^2}{3}] & \text{for } w \leq 1 \\ t_0 + 2t_a + \frac{H}{v_y}[1 + \frac{1}{3w^2}] & \text{for } w > 1 \end{cases} \tag{3.39}$$

If now, T in equation 3.38 is replaced by $\frac{L_x}{v_x}$ the results correspond to those of Gudehus (1972d).

Derivation of the Dual Command Cycle

To determine the dual command cycle (DC), the same assumptions as before for SC with the same modeling of the rack apply. The dual command cycle (DC) consists of two one-way travels from the I/O point to a randomly chosen position in the storage rack and one travel between (TB) those positions. The normalized and dimensionless travel time is:

$$E(DC)^N = 2E(SW_1) + E(TB_1) = E(SC)^N + E(TB_1) \tag{3.40}$$

This is graphically represented in Figure 3.5. To determine $E(DC)$, **the expected travel time between two randomly selected positions** $P_1 = (x_1, y_1)$ and $P_2 = (x_2, y_2)$ is needed, referred to as $E(TB_1)$.

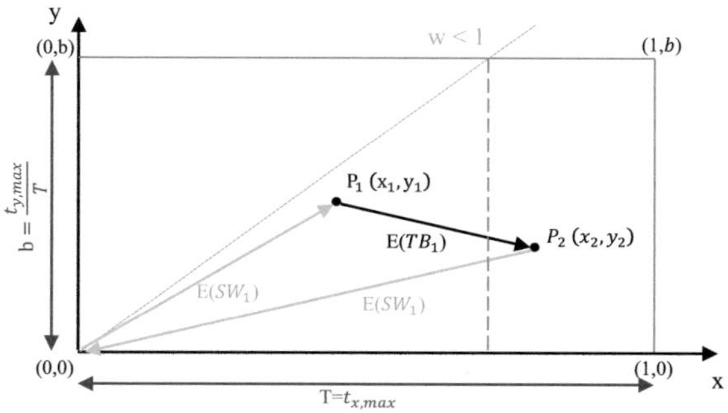

Figure 3.5: Travel time composition of the dual command cycle with two random positions $P_1 = (x_1, y_1)$ and $P_2 = (x_2, y_2)$ according to Bozer and White.

The probability that this time is less than or equal ζ is (Bozer and White 1984, p. 332):

$$Q(\zeta) = P(TB \leq \zeta) = P(|x_1 - x_2| \leq \zeta) \cdot P(|y_1 - y_2| \leq \zeta) \tag{3.41}$$

At first, look at $P(|y_1 - y_2| \leq \zeta)$ and let $f(y)$ and $F(y)$ be the pdf and cdf of the population $y_1, ..., y_n$, respectively. $y_{(1)}, ..., y_{(n)}$ are the order statistics of the sample with sample range $R = y_{(n)} - y_{(1)}$ and sample size n. Based on the random storage policy we have (Bozer and White 1984, p. 332):

$$f(y) = \begin{cases} \frac{1}{b} & \text{for } 0 \leq y \leq b \\ 0 & \text{otherwise} \end{cases} \tag{3.42}$$

and

$$F(y) = \begin{cases} 0 & \text{for } y < 0 \\ \frac{y}{b} & \text{for } 0 \leq y \leq b \\ 1 & \text{for } y > b \end{cases} \tag{3.43}$$

Equations 3.42 and 3.43 are split up because y is limited to the range between 0 and b. $f(y)$ and $F(y)$ are uniformly distributed because of the randomized selection. Let $H(r) = P(R \leq r)$ denote the cdf of the sample range, R, and $h(r)$ the pdf after differentiation. Using equation 3.6 the pdf $h(r)$ can be determined for $n = 2$. Lower and upper bound of the integral result from equation 3.42 (Bozer and White 1984, p. 332).

$$h(r) = 2\int_{v=0}^{b-r} f(v)f(v+r)dv = \frac{2}{b^2}(b-r) \tag{3.44}$$

Let $Q_y(\zeta) := P(|y_1 - y_2| \leq \zeta) = P(R \leq \zeta)$. It holds:

$$Q_y(\zeta) = P(0 \leq R \leq \zeta) = \frac{2}{b^2}\int_0^{\zeta}(b-r)dr \tag{3.45}$$

Solving the integral yields for the y-dimension:

$$Q_y(\zeta) = \begin{cases} \frac{2\zeta}{b} - \frac{\zeta^2}{b^2} & \text{for } 0 \leq \zeta \leq b \\ 1 & \text{for } b < \zeta \leq 1 \end{cases} \tag{3.46}$$

Now consider $P(|x_1 - x_2| \leq \zeta)$ and let $Q_x(\zeta) := P(|x_1 - x_2| \leq \zeta)$. With $0 <= |x_1 - x_2| <= 1$, the probability can be derived analogously to $Q_y(\zeta)$. This yields:

$$Q_x(\zeta) = 2\zeta - \zeta^2, \text{ for } 0 \leq \zeta \leq 1 \tag{3.47}$$

According to equation 3.41, $Q(\zeta)$ can be computed as:

$$Q(\zeta) = \begin{cases} (2\zeta - \zeta^2)(\frac{2\zeta}{b} - \frac{\zeta^2}{b^2}) & \text{for } 0 \leq \zeta \leq b \\ 2\zeta - \zeta^2 & \text{for } b < \zeta \leq 1 \end{cases} \tag{3.48}$$

The derivative of $Q(\zeta)$ is

$$q(\zeta) = \begin{cases} (2 - 2\zeta)(\frac{2\zeta}{b} - \frac{\zeta^2}{b^2}) + (2\zeta - \zeta^2)(\frac{2}{b} - \frac{2\zeta}{b^2}) & \text{for } 0 \leq \zeta \leq b \\ 2 - 2\zeta & \text{for } b < \zeta \leq 1 \end{cases} \tag{3.49}$$

When integrating over ζ, we obtain the mean travel between distance

$$E(TB_1) = \int_0^1 \zeta q(\zeta) d\zeta = \frac{1}{3} + \frac{1}{6} b^2 - \frac{1}{30} b^3 \tag{3.50}$$

For $b = 1$ the normalized travel-between distance is $\frac{14}{30} = \frac{7}{15}$. According to equation 3.40, the dual command cycle is obtained. Again, the result needs to be re-scaled by multiplying with T.

$$E(DC) = (\frac{4}{3} + \frac{b^2}{2} - \frac{b^3}{30})\ T \tag{3.51}$$

Gudehus (1972d) present the following result for expected mean dual command cycles time with w and t_a as in equations 3.19 and 3.21:

$$E(DC) = \begin{cases} t_0 + 3t_a + \frac{L}{v_x}[\frac{4}{3} + \frac{1}{2}b^2 - \frac{1}{30}b^3] & \text{for } w \leq 1 \\ t_0 + 3t_a + \frac{H}{v_y}[\frac{4}{3} + \frac{1}{2b^2} - \frac{1}{30b^3}] & \text{for } w > 1 \end{cases} \tag{3.52}$$

Again, if T in equation 3.51 is replaced by $\frac{L_x}{v_x}$ the results correspond to those of Gudehus (1972d).

Travel Time Formulas allowing for dwell times

Allowing for all actual amounts of time a command cycle consists of, the mean travel time for single and dual command cycle in detail for $b = 1$ are formulated as follows:

$$E(SC) = 4t_{dead} + 2t_{mast} + 2t_{LHD} + (\frac{v_x}{a_x} + \frac{v_y}{a_y}) + \frac{4}{3}T \tag{3.53}$$

$$E(DC) = 8t_{dead} + 3t_{mast} + 4t_{LHD} + \frac{3}{2}(\frac{v_x}{a_x} + \frac{v_y}{a_y}) + \frac{9}{5}T \tag{3.54}$$

Official Guidelines

Because of its practical relevance, the Association of German Engineers (VDI) and the European Federation of Materials Handling (FEM) published guidelines for travel time determination of AS/RSs. Both do not present travel time models for mathematical calculation, but define representative test cycles. The idea is to define a mean 'representative cycle' that is in accordance with the results of the mathematically defined travel time models. By performing that cycle multiple times with an AS/RS and measuring the time needed, both performance and throughput of the system are determined.

To compose this cycle, representative positions that lead to the mean travel time, are indicated. These positions can be derived from the results of Gudehus (1972d) and Bozer and White (1984) for $b = 1$. From equation 3.53 follows, that the coordinates of those representative positions for the single command cycle are located at $\frac{2}{3}$ of the height or length of the rack. With $\frac{L}{v_x} = \frac{H}{v_y} = T$ and equation 3.20, we can define many positions having the path-depending travel time of a single command cycle. They are located on the isochrone with

$$x = \frac{2}{3}L \quad , \quad y = \frac{2}{3}H \tag{3.55}$$

that is also shown in Figure 3.6. Note that all positions lying on that isochrone represent positions for a single command cycle with the mean cycle time.

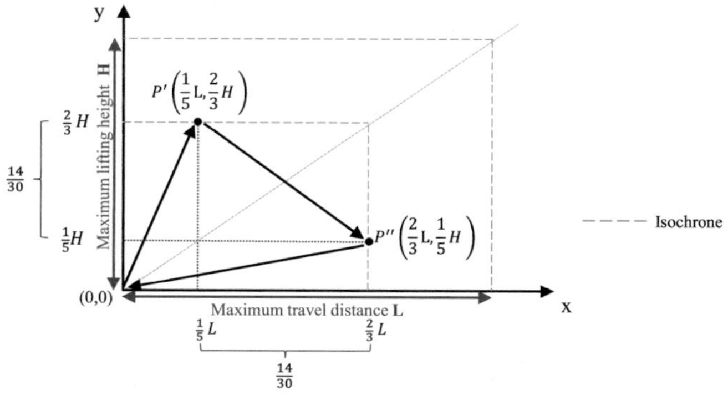

Figure 3.6: Mean dual command cycle with representative positions.

To define the mean dual command cycle, those two positions, that lie at the isochrone and exhibit a travel between distance of $\frac{14}{30}$ (see equation 3.50) are required. Two positions that satisfy these requirements are $P' = (\frac{1}{5}L, \frac{2}{3}H))$ and $P'' = (\frac{2}{3}L, \frac{1}{5}H))$ as depicted in Figure 3.6. Note that P' and P'' show the travel between distance in both directions, thus $|x' - x''| = \frac{14}{30}$ equals $|y' - y''| = \frac{14}{30}$.

Consequently, the mean dual command cycle in FEM 9.851 (1978) is defined by:

$$P_1 = (\frac{1}{5}L, \frac{2}{3}H) \quad , \quad P_2 = (\frac{2}{3}L, \frac{1}{5}H)) \tag{3.56}$$

In VDI Richtlinie 3651 (1973), the travel between distance is approximated with $\frac{1}{2}$ resulting in the following representative positions:

$$P_1 = (\frac{1}{6}L, \frac{2}{3}H) \quad , \quad P_2 = (\frac{2}{3}L, \frac{1}{6}H)) \qquad (3.57)$$

The guidelines point to the fact that the method is sufficient accurate for racks with a shape factor in the range of [0.5;2] (FEM 9.851 1978, p. 6), (VDI Richtlinie 3651 1973, p. 4).

3.2.2 Further Development of the Fundamental Travel Time Model

In the research community, a variety of adaptions and enhancements of the the basic models exist. There are many groups of travel time models focusing on special storage types, special configurations of the rack or on different storage- or operating policies. Cycle times of traditional AS/RSs are determined mathematically for mostly all aspects, whereas for non-traditional systems (e.g. systems for multiple load handling) many open research questions exist (Roodbergen and Vis 2009).

Hwang and Lee (1990) and Chang et al. (1995) consider both speed and acceleration/deceleration of the S/R machine and extend the work of Bozer and White (1984) in this respect. Wen et al. (2001) continue this research under the assumption of class based storage. In addition to Hausman et al. (1976) and Graves et al. (1977), Gudehus (1972a) is among the first to develop analytical models for a storage assignment with two or three classes. Rosenblatt and Eynan (1989) present an approach to define optimal boundaries of n classes in an AS/RS. Based on that, Eynan and Rosenblatt (1994) analyze the influence of the shape factor, number of classes and demand characteristics for one-way travel times. Additional authors study the effect of class based storage, present further travel time models and analyze optimal dimensioning of classes (e.g. Kim and Seidmann (1989),Kouvelis and Papanicolau (1995), Van den Berg (1996), Ashayeri et al. (2002)). Bortolini et al. (2015) provide a comprehensive summary of relevant literature in the context of class based storage. Moreover, they present general analytical models for a single and dual command cycle in a three class based warehouse and perform a sensitivity analysis to support practitioners with advice for optimal class dimensioning depending on both the shape factor and the skewness factor of the demand.

Problems related to routing and sequencing offer an additional research section, as smart sequencing allows throughput improvements compared to the **first come first served** (FCFS) selection of requests (Roodbergen and Vis 2009). Gudehus (1972c) is the first to discuss the idea of finding storage or retrieval positions within the no-cost area to reduce travel times. Han et al. (1987) adopt the idea and denote it as the shortest leg heuristic. Sequencing heuristics for traditional AS/RSs are presented by Eynan and Rosenblatt (1993), Schwarz et al. (1978) and Ascheuer et al. (1998). Lee and Schaeffer (1996), Lee and Schaeffer (1997) and Van den Berg and Gademann (1999) present optimal solutions of the sequencing problem for selected cases.

However, many other types of non-traditional AS/RSs exist, which is why there are a various number of publications dealing with travel time modeling of shuttle based S/RS, person-on-board, channel storage or carousel systems. For a more detailed review of such systems, we refer to the following publications: Sarker and Babu (1995) as well as Vasili, Tang and Vasili (2012) provide literature reviews concerning design, operation and performance evaluation of AS/RS. Johnson and Brandeau (1996) focus their review on analytical-stochastical models in the context of AS/RS design and control. Roodbergen and Vis (2009) present an extensive study in the area of AS/RS planning and control with respect to warehouse design, storage assignment and operating policies, such as routing, sequencing and batching. Gu et al. (2010) consider also warehouse design issues and case studies.Gagliardi, Renaud and Ruiz (2012) provide a review that contrasts statistical and stimulative approaches of AS/RS performance evaluation studies. A recent review that also incorporates shuttle based systems is presented by Kalyanaraman and Keerthika (2016).

3.3 Related Work

In this section, we address existing literature in the field of AS/RS performance evaluation that is relevant for our own approach. We distinguish between three different groups of related work:

1. Approaches for routing and sequencing that provide starting points for the strategies we define.
2. The travel time model for double deep AS/RSs our model is based on.
3. Travel time consideration for double deep, dual capacity load handling AS/RSs that have limitations for different reasons.

Each subsections in the following addresses one of the groups.

3.3.1 Routing and Sequencing approaches

The work of **Han et al. (1987)** for dual command cycles in traditional AS/RSs is a frequently referenced study that provides the basis for the work of many other researchers. The authors introduce the **Nearest Neighbor** heuristic that sequences the nearest retrieval position to an open location with the objective to reduce the travel between distance in a dual command cycle. The objective is to select an efficient pair of storage and retrieval positions. In a first step, they derive the pdf for the smallest of k randomly chosen distances, using the results of Bozer and White (1984), who present pdf and cdf for the distance between two randomly selected positions. The smallest of k randomly chosen distance is a random variable, Z_k, with the pdf $r(Z_k)$.

$$r(Z_k) = k[1 - Q(Z_k)]^{k-1} q(Z_k) \text{ for } \leq Z_k \leq 1 \tag{3.58}$$

$Q(\zeta)$ and $q(\zeta)$ are known from equations 3.48 and 3.49. The expected smallest distance is (Han et al. 1987, p. 59):

$$E(Z_k) = \int_0^1 \zeta k[1 - Q(\zeta)]^{k-1} q(\zeta) d\zeta \tag{3.59}$$

Based on this result, the authors present an approximation of the travel between distance for m open locations and a block of n requested retrieval positions:

$$E(TB_{n,m}^{NN}) = \frac{1}{n} \sum_{i=m}^{n+m-1} E(Z_k) \tag{3.60}$$

For the analytical formulation of the proposed dual command cycle, $E(TB)$ is substituted with $E(TB_{n,m}^{NN})$ in expression 3.40.

For a block size of 20 retrieval requests they report increased throughput by 18% for the dual command nearest neighbor policy with one open location. The throughput can be increased further with a greater number of open locations.

Next, the authors formulate the **Shortest Leg** heuristic in order to find a lower bound of the DC under the nearest neighbor heuristic. The Shortest Leg heuristic selects storage and retrieval position from m open locations and n retrieval positions that create the least total travel distance between the I/O point and the retrieval position. This means, storage positions from the no-cost zone are selected, if possible. Analytical results show that throughput is improved further when applying the shortest leg heuristic. Using Monte Carlo simulation, the dynamic behavior of the heuristic is studied. They find that the Shortest Leg heuristic changes the distribution of open locations by moving the open locations away from the I/O point. On the contrary, the nearest neighbor heuristic shows a constant performance and therefore outperforms Shortest Leg on the long run.

Note that equation 3.59 is also denoted as the mean travel between distance between one randomly selected position and the nearest of m randomly selected positions, or $E(TB_m)$.

Sarker et al. (1991), Sarker et al. (1994) and Keserla and Peters (1994) present similar approaches suggesting a quadruple command cycle that is executed according to the **Flip Flop** policy. Based on the approach of Bozer and White (1984), they present an analytical formulation of the corresponding cycle time. Sarker et al. (1991) adjuste the cycle by minimizing the travel between distances based on the the nearest neighbor idea, i.e., a storage and a retrieval position are chosen near to the position of the Flip Flop operation. As proposed in Han, McGinnis, Shieh and White (1987), they formulate a lower bound of the travel time by selecting a storage and a retrieval position from the no-cost zone. Simulation is used to validate the formulated heuristics and compare the performance to single load handling. They report improvements ranging from 50% to 80% and recommend dual load handling systems. Based on these findings, Sarker et al. (1994) analyze the previous system for class based storage. They present a heuristics for a quadruple command cycle on a two class basis and assess their analytical model with a simulation. With both methods, they report possible throughput improvements of up to 25% in comparison to the results without class based storage. Keserla and Peters (1994) also combine the nearest neighbor idea with the Flip Flop heuristic, but adjust this combination in such a way that the minimum perimeter of the triangle, defined by the three stops of the cycle, is chosen. Upper and lower bounds of the heuristic based on the ideas of Han, McGinnis, Shieh and White (1987) are presented. They evaluate the heuristic via Monte Carlo simulation and show a throughput improvement of 25% compared to a nearest neighbor dual command cycle. This is a lower potential compared to the results of Sarker et al. (1991), which can be explained by the allowance for load handling times.

Meller and Mungwattana (1997) present three sequencing heuristics, both for quadruple and sextuple command cycles. Based on the results of Bozer and White (1984), they use order statistics to derive expected smallest and expected largest one-way travel time from the I/O point to one of m randomly selected locations, $E(SW_m)$ and $E(SL_m)$, respectively. Using equations 3.34 and 3.35, they derive:

$$E(SW_m) = \int_0^1 \zeta m[1 - G(\zeta)]^{m-1} g(\zeta) d\zeta \tag{3.61}$$

$$E(SL_m) = \int_0^1 \zeta m[G(\zeta)]^{m-1} g(\zeta) d\zeta \tag{3.62}$$

All heuristics are based on the assumption of a fixed number of retrieval position (i.e., two or three) and m open locations. The first nearest neighbor heuristic composes the cycle in the following way: Nearest storage location — second nearest storage location — third nearest storage location — nearest retrieval position — second nearest retrieval position — third nearest retrieval position — return. The quadruple command cycle is composed accordingly. In a similar variant, the Reverse Nearest Neighbor heuristic (RNN), the retrieval (storage) location closer to the I/O point is approached lastly (first). In their so called modified command cycle, they combine the Nearest Neighbor principle with the Flip Flop heuristic. Stops within a cycle that are not involved in Flip Flop are arranged in the same way as before. Using the newly derived formulas and the Nearest Neighbor travel between distance from Han et al. (1987) (equation 3.60), travel time estimates are presented. Expected travel times are numerically evaluated for different examples of m. They report that the modified quadruple command cycle in combination with the nearest neighbor idea performs about 36% better than the dual command cycle under nearest neighbor policy, for m ranging between 1 and 10.

Eynan and Rosenblatt (1993) analyze the application of the Nearest Neighbor idea in a class-based storage environment by selecting storage and retrieval pairs from the same class. They derive the mean travel between distance for Nearest Neighbor selection in each of i classes and the total expected travel between distance as the average from all classes. They report a reduced travel between distance of up to 65% compared to Han et al. (1987) for six classes.

Grafe (1997) presents a qualitative consideration of AS/RS with multiple load handling and addresses possible advantages of a Flip Flop policy and class based storage. No mathematical travel time models are formulated. Based on a rule of thumb that approximates the travel distance in x-direction, a method for throughput determination is presented.

Potrč, Lerher, Kramberger and Šraml (2004) present an approach to perform quadruple and sextuple command cycles according to a heuristic called 'Strategy x'. In this heuristic, they propose to randomly choose storage and retrieval positions, which are sequenced in an ascending order in x-direction of the rack. Performance is evaluated by a simulation model in which different system configurations, consisting of rack dimension and velocity profiles, are simulated. In this way, they show improved throughput potential of multiple load handling systems on the one hand and dependencies of travel times on rack dimensions and velocities on the other hand.

Kraul (2010) considers performance models for different kinds of AS/RSs. He presents an adjusted version of the Shortest Leg heuristic for multiple-load handling devices to overcome the reported drawbacks of Han et al. (1987). He proposes to chose a random storage location first, before locations from the no-cost zone are selected. In this way, the shift of available positions away from the I/O point is prevented.

3.3.2 Travel time models for double deep AS/RS

Lippolt (2003) is the first to develop an analytical model for double deep AS/RSs with an exact determination of the rearrangement probability. The author describes the operation of the storage place as a stochastic process where the storage and retrieval events form a Markov Chain. The process is stationary which allows to derive the stationary distribution of the storage lanes, i.e., the mean number of empty, half-filled and filled storage lanes as a function of the filling level. Based on the stationary distribution, he studies all possible retrieval operations and the occurrence of rearrangements (R). With z being the filling level, the probability that a rearrangement occurs is:

$$P(R) = \frac{z}{1+z} \tag{3.63}$$

Next, the mean access times of the load handling device are determined, as they also depend on the stationary allocation of a storage lane.

Additionally, a closed form approximation to determine the expected distance of the nearest rearrangement position ($E(UF)$) is developed. He is the first to incorporate the total number of storage positions and their occupancy state in addition to the filling level. A closed form approximation for the closest distance out of k random positions is developed, which is:

$$E(UF) = \frac{7}{15}\frac{1}{\sqrt{k}} \tag{3.64}$$

Equation 3.64 converges to zero for $k \to \infty$, although in reality, the **minimum distance** to a rearrangement positions **is one**. Therefore, the approximation is corrected in order to model a discrete rack and improve the approximation quality. This is:

$$E(UF) = (\frac{7}{15})^{(1-\frac{k}{l})}\frac{1}{\sqrt{k}} \tag{3.65}$$

where l is the total number of storage lanes. When k increases, the influence of $\frac{7}{15}$ in equation 3.65 decreases. For l = k, i.e., all storage lanes serve as rearrangement positions, the **rearrangement distance is one**.

He concludes that the total rearrangement effort does not extend mean travel time as feared and that also random storage allocation within the storage lanes is economically reasonable (Lippolt 2003, p. 161f)

3.3.3 Dual Load Handling combined with Double Deep Storage

There are some publications which discuss double deep AS/RSs with a dual capacity load handling devices. All publications, as far as we know, are presented in the following. They considerably differ from each other because of their focus of investigation and approach.

Ritonja (2003) and Oser and Ritonja (2004) study two specific AS/RSs in the context of class based storage: One is a single deep rack operated by an S/R machine with a fourfold load handling device with two loads on top of each other. The second one is an double deep AS/RS with two double deep load handling devices and in consequence with a double deep aisle. For three different types of class based storage assignments, they formulate general analytical models for multiple command cycles. Subsequently, they simulate the operation of both AS/RS configurations in a buffer stock system where storage and retrieval requests are performed at separate times.

Kayser (2003) presents an application of the results of Lippolt (2003) for the dual load system with two LHD next to each other. He formulates the following assumptions for the quadruple command cycle: The proposed cycle is always executed in the order storage-storage-retrieval-retrieval. If a rearrangement is required at the position of the first retrieval, both units are picked up with the load handling devices. The blocked unit and the retrieval unit are carried to the rearrangement position. The S/R machine moves immediately from the rearrangement position to the second retrieval position. The rearrangement probability and the access times of the load handling device are adopted from Lippolt (2003).

Seemüller (2006) defines several analytical models for miniload AS/RSs that also involve dual load handling in a double deep storage environment. In all configurations, he examines FCFS, a Nearest Neighbor policy that selects storage positions near to given retrieval positions and Multiple Storage policies. The approach which is to define the components of the respective travel times, has the following shortcomings. For double deep storage, he applies the rearrangement probability from Lippolt (2003). In contrast to Lippolt (2003), the rearrangement distance is simplified: The rearrangement distance is set to one (i.e., the adjacent storage position is available), for all filling levels below 95%. For higher filling levels, he uses a general limit value. The same simplification is applied for the proposed Nearest Neighbor policy. The possibility of using the two load handling devices to re-store a rearrangement unit into the same storage is not considered. Moreover, the different execution orders of the quadruple and sextuple command cycle are not regarded. In an calculation example for double deep storage with triple load handling, he finds the Multiple Storage policy to perform best up to filling level of 75%. Above that, it is outperformed by the nearest neighbor policy.

Xu et al. (2015) present analytical travel time models for a quadruple command cycle operating under both a FCFS and a Nearest Neighbor policy. They also assume the execution order being storage-storage-retrieval-retrieval, always. Rearrangements are performed either by means of the load handling device, if a rearrangement for the first retrieval operation occurs, or by relocating the blocking unit to the closest available storage lane. In their approach, they define nine different cases in which a cycle can be performed and derive a weighted average to determine the cycle time. The allocation of the rack is simplified in two ways:

- In the first model, the front positions are used only if all storage lanes are occupied at the rear position as seen in Lerher et al. (2010), Garlock (1997) and Ritonja (2003).
- In the modified model, empty storage lanes above a filling level of 50% are allowed and, if possible, a single storage lane is used for both storage operations. In the associated analytical model, the number of half-filled storage lanes is required as an additional input parameter. This means, the exact state of the storage lanes needs to be known to calculate the cycle time. Especially during the dimensioning of an AS/RS, these states can only be assumed.

Moreover, the mean rearrangement cycle time is not specified. Further, a mean rearrangement distance is not defined. The authors find travel times of the FCFS quadruple command cycle to outperform the dual command cycle by at least 21%. Under the nearest neighbor policy, reported travel times can be improved by additional 9% to 12%.

3.3.4 Conclusion on Literature for Travel Time Determination and Derivation of the Research Questions

We learn from literature that throughput can be improved considerably when applying quadruple command cycles instead of dual command cycles (e.g. Gudehus (1972e),Meller and Mungwattana (1997)). Lippolt (2003) derives the additional travel time due to rearrangements in a mean dual command cycle. He shows that double deep storage can represent an efficient alternative to usual, single deep storage. However, little attention is given to the configuration that combines quadruple command cycles with double deep storage environments. Existing studies are characterized by limitations in terms of general travel time modeling, as they do not allow for totally randomized storage allocation, all possible orders of execution and the different possibilities for rearrangement. Moreover, they have constraints in approximated components of the travel time formulation (Seemüller 2006) or impractical input requirements (Xu et al. 2015). The particular stationary allocation of the storage lanes relevant for double deep storage as derived by Lippolt (2003) is not taken into account. For the above reasons, we propose our first research question:

How can the mean travel time of a randomly executed quadruple command cycle be accurately determined?

In Chapter 4 we answer this question by formulating a general travel time model that overcomes the current restrictions.

It is commonly observed that using routing and sequencing methods can significantly decrease travel times in AS/RSs, regardless of their specific configurations. As a result, many researches have applied the methods of Han et al. (1987) such as the combined usage of the Flip Flop idea with no-cost or Nearest Neighbor selection of requests for quadruple command cycles proposed by Sarker et al. (1991). Recently, authors considering the double deep, dual load handling case propose the application of Nearest Neighbor based heuristics or Multiple-Storage (Xu et al. (2015)), but do not succeed to provide all suitable routing and sequencing methods this configuration allows. Moreover, a detailed investigation of how the rearrangement behavior can be influenced using the dual load handling device has not been addressed yet. Consequently, an investigation of all possible operating strategies and their evaluation is targeted by our research question two:

How can sophisticated operating strategies be described and in which cases do they provide a travel time advantage?

We cover the second question by setting up various routing and sequencing strategies for the execution of a quadruple command cycle in Chapter 5, before we further evaluate selected strategies with a simulation model in Chapter 6.

4 Analytical Models for Dual Load Handling, Double Deep AS/RSs

Each equation in a book
would halve the sales.
-S. Hawking

This chapter introduces the mathematical model for determining the travel time of the quadruple command cycle (QC) in the context of double deep storage. We choose a stochastic approach to account for double deep storage in combination with dual load handling based on Lippolt (2003). Storage, retrieval and rearrangement operations within a quadruple command cycle are modeled as a stochastic process. The objective is to determine the rearrangement effort which is an essential part of the travel time formula.

We derive a general version of the model and propose an additional variant emerging from the rearrangement options. The validity of our approach is shown by application of a simulation model. Parts of this Chapter have been published by Dörr and Furmans (2016b) and Dörr and Furmans (2016a).

4.1 Basics for Modeling a Quadruple Command Cycle

The modeling setup is a single storage aisle with double deep storage racks on both sides, operated by an automated S/R machine with a dual capacity load handling device (LHD).

4.1.1 Assumptions

In our analytical model, we consider only one side of the aisle and make the following assumptions:

Rack shape

The rack is rectangular and has a fixed number of storage lanes; each lane consists of two storage positions. This means the number of **storage positions** (l^*) at one side of the rack is:

$$l^* = l_{horizontal} \cdot l_{vertical} \cdot 2 \tag{4.1}$$

$l_{horizontal}$ and $l_{vertical}$ are the number of lanes in horizontal and vertical direction, respectively. The **number of storage lanes** is $l = l^*/2$. We assume the rack to be square-in-time, i.e., with a shape factor of 1.

I/O position

The I/O point is located at the bottom left corner of the rack.

Dwell times

We include a term for dwell times that is added to the path-depending travel times. Unless stated otherwise, we assume the following:

- One t_{mast} per movement of the S/R machine, thus per travel between different positions.
- Two t_{dead} per access cycle of the load handling device.

Storage policy

A random storage assignment policy is applied. For retrieval, any occupied position has an equal probability to be selected. Therefore, in a fully occupied storage lane, both units have the same selection probability. For storage, any storage lane that is not fully occupied has an equal selection probability. Units are always stored in the rearmost position, which is why it is equally likely to choose an empty storage lane or a half-filled storage lane. Since the lanes are selected randomly, fully occupied lanes can also exist below filling levels of 50%.

Storage lane allocation

Storage lanes that are only occupied in the front position are not possible. This remains valid during rearrangements. The state *half-filled front* in Figure 4.1 is excluded in our model. For practical applications, this state would only make sense for low filling levels to save handling time. Otherwise storage space in the rear positions would be blocked.

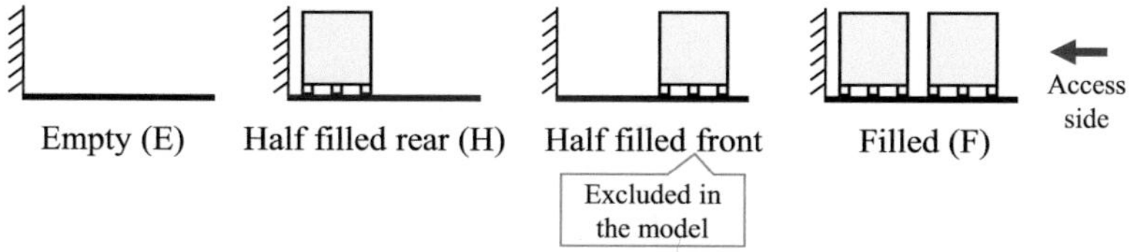

Figure 4.1: All potential states of a storage lane

The empty state of a storage lane is denoted as **E**, the half-filled state as **H** and the filled state as **F**.

Load handling

The two load handling devices can handle goods independently and are both able to access the front and the rear position of a storage lane. We distinguish between the following access times:

- $t_{LHD,f}$: The access time to the **front** position, meaning the time the load handling device needs to extend to the front position of a lane and move back to the initial position. We assume the handling time at the I/O point equivalent to the access time to the front position.
- $t_{LHD,r}$: The access time to the **rear** position.

The devices are horizontally arranged. There is a simultaneous process at the beginning and the end of every command cycle for pick-up and deposit of units at the I/O point. The distance between the two load handling devices is equivalent to the distance between two storage lanes. Hence, the positioning of one load handling device in front of one storage lane means the other load handling device is necessarily positioned in front of a neighboring lane. We discuss the impacts for the relaxation of this assumption in section 4.6.1.

The utilization order of the load handling devices is randomly chosen.

Rearrangements

For each retrieval of units stored at the rear position of a storage lane, there is a positive probability that a rearrangement is required. Rearrangements occur, if a unit at the rear position is retrieved, while the front position is occupied. Storing in the rearmost position also applies during rearrangements.

We distinguish between **regular** rearrangements and **tango** rearrangements:

- **Regular** rearrangements describe the process of rearranging the blocking unit into an available storage position. Subsequently the retrieval unit is handled. Regular rearrangements are conducted in case that one handling device is in use. For the selection of the rearrangement position, the Nearest Neighbor policy applies.

- The **tango** rearrangement is a combined operation of retrieval and rearrangement. Every time both load handling devices are free and a rearrangement is required, it is possible to perform a tango which works as follows: The blocking unit in the front is picked by the first load handling device (Figure 4.2 a). In the second step, the S/R machine moves sideways to bring the other load handling device in front of the same storage lane. Then, the (formerly blocked) unit in the rear is picked by the second load handling device (b). At last, the S/R machine moves back and the blocking unit is restored to the rear position of the same storage lane (c). In case a tango is performed, no additional storage position for rearrangement is required. The tango rearrangement is performed whenever possible. The order in which the load handling devices are used is randomly selected.

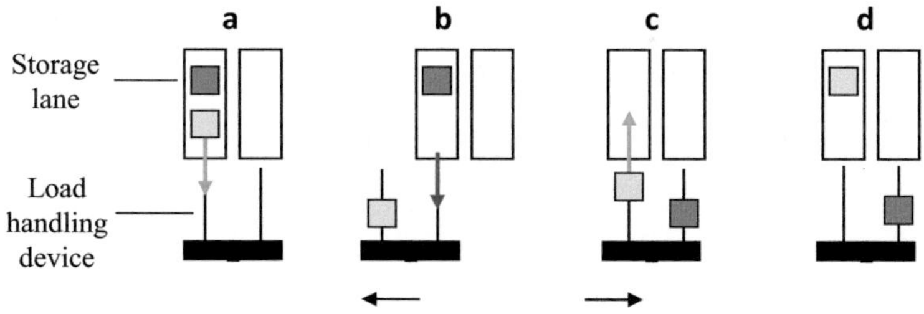

Figure 4.2: Rearrangement performed by the load handling devices (**tango**)

System load

We consider the AS/RS to be operated at full capacity, meaning there are always storage and retrieval requests waiting. Consequently, the AS/RS performs only quadruple command cycles.

Filling Level

The filling level, z, of the whole rack depends directly on the condition of the storage lanes. Thus, the filling level can be calculated given the number of half-filled and filled storage lanes in the following way:

$$\begin{aligned} z &= \frac{1/2 \cdot \text{Number of half-filled lanes} + \text{Number of filled lanes}}{\text{Total number of lanes}} \\ &= 1/2 \cdot P(H) + P(F) \end{aligned} \tag{4.2}$$

We refer to the filling level as z.

4.1.2 Role of the Execution Order

A quadruple command cycle consists of two storage and two retrieval operations. Each cycle begins with the pick-up of two storage units and ends with the deposition of two retrieved units.

Possible execution orders

The locations within a cycle are randomly selected and approached one after another. A retrieval is only possible if at least one handling device is available. This means each retrieval operation needs one preceding storage operation and the QC can be executed in two different orders:

- Storage - Storage - Retrieval - Retrieval (**SSRR**)
- Storage - Retrieval - Storage - Retrieval (**SRSR**)

Generally, both orders are possible. The probability that the cycle is executed according to the SSRR order is $P(SSRR)$. The probability for the SRSR order is $P(SRSR) = 1 - P(SSRR)$. Figure 4.3 illustrates the procedure during both execution orders.

In the case of SSRR, the S/R machine stores both units first, resulting in a completely free load handling device. If the first retrieval unit is blocked and thus a rearrangement is needed, a **tango** rearrangement is always performed. In all other cases, **regular rearrangements** are performed.

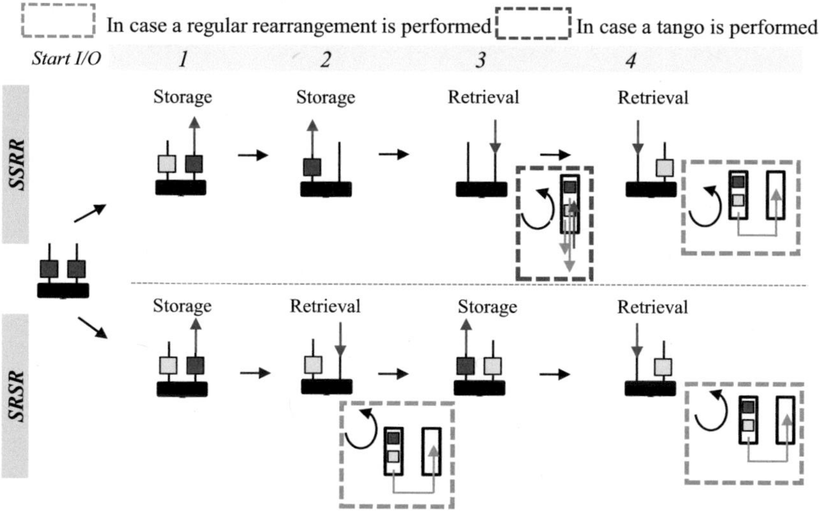

Figure 4.3: Graphic illustration of the two possible execution orders

In practice, the probability of both execution orders ($P(SSRR)$ and $P(SRSR)$) may be determined through the respective control system or a particular routing strategy. In this chapter, we use the execution order as variable to obtain a general solution in this regard. For evaluation purposes, we apply three different realizations of the probability distribution of the execution order which are shown in Table 4.1. This allows us to discuss which execution order performs best. In subsequent chapters, the execution order is not always variable, but often determined by the routing strategy that is applied.

Probability distribution of the execution order	Description
$P(SSRR;SRSR) = (\frac{1}{2};\frac{1}{2})$	Basic model
$P(SSRR;SRSR) = (1;0)$	SSRR model
$P(SSRR;SRSR) = (0;1)$	SRSR model

Table 4.1: Calculated execution orders presented in this thesis

The **basic model** refers to the idea that the execution order is chosen randomly in order to model a totally random base case, hence both execution orders are equally likely ($P(SSRR; SRSR) = (\frac{1}{2}; \frac{1}{2})$). In the SSRR model, the S/R machine only operates in the SSRR order while in the SRSR model, only the latter mentioned order is applied.

Analytical composition incorporating the execution Order

Analytically, a quadruple command cycle consists of a single command cycle and three travel between distances (Meller and Mungwattana 1997). The normalized and dimensionless (indicated by the superscript N) travel time therefore is:

$$E(QC)^N = E(SC)^N + 3E(TB) \tag{4.3}$$

The execution order does not influence equation 4.3 and thus is not part of this expression. Due to double deep storage, rearrangements can occur with every retrieval and the extra time caused by rearrangements needs to be added.

However, the occurrence of tango and regular rearrangement is dependent on the execution order, as SSRR allows a tango, while SRSR does not (see Figure 4.3). Performing the cycle in the SSRR order generates a different probability that a rearrangement occurs compared to the SRSR order (the reason is a different allocation of the storage lanes and is discussed in further detail in the following section). $P^{SSRR}_{Rearrange}$ and $P^{SRSR}_{Rearrange}$ give the rearrangement probabilities for the respective execution order and thus are different from each other. $E(RC)$ is the expected mean cycle time of a regular rearrangement command, t_{Tango} the time needed for a tango rearrangement. Since every cycle consists of two independent retrieval operations, we find both $P^{SSRR}_{Rearrange}$ and $P^{SRSR}_{Rearrange}$ twice for every execution in the analytical formulation. We adjust equation 4.3 accordingly to allow for rearrangements in double deep ($E(QC)_{dd}$) storage environments:

$$
\begin{aligned}
E(QC)_{dd} = & E(SC)^N + 3E(TB) \\
& + P(SSRR)(P^{SSRR}_{Rearrange} \cdot t_{Tango} + P^{SSRR}_{Rearrange} \cdot E(RC)) \\
& + P(SRSR)(2 \cdot P^{SRSR}_{Rearrange} \cdot E(RC))
\end{aligned}
\tag{4.4}
$$

A model that consists of different rearrangement probabilities, i.e., $P^{SSRR}_{Rearrange}$ and $P^{SRSR}_{Rearrange}$, is inconvenient. We want to define a mean rearrangement probability $P_{Rearrange}$, that is valid for both orders in a way that we can rewrite equation 4.4:

$$
\begin{aligned}
E(QC)_{dd} = & E(QC)^N \\
& + 2 \underbrace{1/2 P(SSRR) P_{Rearrange}}_{P(Tango)} \cdot t_{Tango} \\
& + 2 \underbrace{(1 - 1/2 P(SSRR)) P_{Rearrange}}_{P(Regular)} \cdot E(RC))
\end{aligned}
\tag{4.5}
$$

$(1 - 1/2 P(SSRR))$ and $1/2 P(SSRR)$ are the conditional probabilities of a regular rearrangement respectively a tango, given a rearrangement takes place. $P(Regular)$ and $P(Tango)$ are the unconditional probabilities that, for any execution order, a regular rearrangement respectively a tango occurs. We will further investigate these relations in the next section.

4.2 Approach for the General Model of a Random Quadruple Command Cycle

The first objective of this section is to determine the rearrangement probability, $P_{Rearrange}$, considering the possible execution orders of a QC. We aim for a general solution and thus present a formulation with $P(SSRR)$ as a variable.

Intuitively, the rearrangement probability is depending on the filling level z. With an increasing number of occupied storage positions, the likelihood that a blocked unit is selected, increases. However, the filling level itself is not sufficient to determine the rearrangement probability, as a certain filling level can be represented by different allocations of the storage lanes. The storage lane allocation corresponds to the percentage of empty, half-filled and filled storage lanes. The probability that a storage lane is in state i is given by $P(i), i \in \{E, H, F\}$.

The stored units are not equally distributed over the front and rear position. The reason is the assumption that half-filled storage lanes are only occupied at the rear position, while storage lanes occupied at the front position do not exist (see Figure 4.1). As the different storage, retrieval and rearrangement operations have immediate impact on the exact storage lane allocation, we need to analyze them individually.

The steps in order to define $P_{Rearrange}$ in subsection 4.2.1 are:

1. We model the **allocation of a storage lane as a stochastic process**.
2. Its stationary distribution yields the **probability distribution of the storage lane allocation** as a function of the filling level.
3. We use this probability distribution to **derive the rearrangement probability**.

Following this procedure, we are able to determine **mean load handling times** for storage and retrieval operations and **E(RC)**, which are depending on the probability distribution, in subsection 4.2.2.

The results, consolidated in subsection 4.2.3, can be customized by applying them to any particular probability distributions of the execution orders $P(SSRR; SRSR)$. We present numeric calculation for three examples of the probability distribution in section 4.3 (see Table 4.1).

4.2.1 Rearrangement - Probability

In this subsection, the rearrangement probability is derived according to the steps mentioned at the beginning of this section. In order to model a stochastic process, all possible storage and retrieval events are considered next.

Possible storage and retrieval events

The storage lanes change their states during storage and retrieval. We investigate how storage or retrieval operations can affect the storage lane allocation and determine the probabilities of these events.

Storage Operations All storage lanes that are empty or half-filled provide the possibility of storage. Consequently, there are two possible storage operations. The first possibility is storing into an empty storage lane (**S1**), which represents the transition from an empty storage lane to a half-filled storage lane. The second possibility describes storing into a half-filled storage lane (**S2**), which results in the change from a half-filled storage lane to a fully occupied storage lane. Both transitions are shown in Figure 4.4.

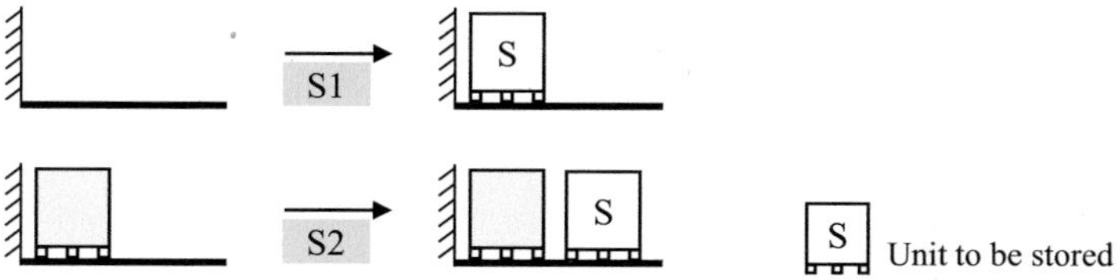

Figure 4.4: Possible storage events

The probability of S1, $P(S1)$, represents the probability of choosing an empty storage lane from all half-filled or empty storage lanes. At the same time, it is the probability of choosing an empty storage lane (E) given a storage operation (S) takes place. Equally, $P(S2)$ represents the probability of choosing a half-filled lane for storing.

The probabilities of storing into an empty or a half-filled lane can be specified as follows:

$$P(S1) = P(E|S) = \frac{P(E)}{P(E) + P(H)} \tag{4.6}$$

$$P(S2) = P(H|S) = \frac{P(H)}{P(E) + P(H)} = 1 - P(S1) \tag{4.7}$$

Retrieval Operations We distinguish five retrieval operations: Two operations without rearrangement, R1 and R2, two retrieval operations with a regular rearrangement, R3 and R4, and the tango, represented by R5.

If a unit is retrieved from a half-filled storage lane, we refer to this as **R1**. In this case, the storage lane changes from the half-filled state to the empty state. Retrieving a front unit from a filled storage lane, resulting in a half-filled storage lane, is denoted as **R2**. The transitions are shown in Figure 4.5.

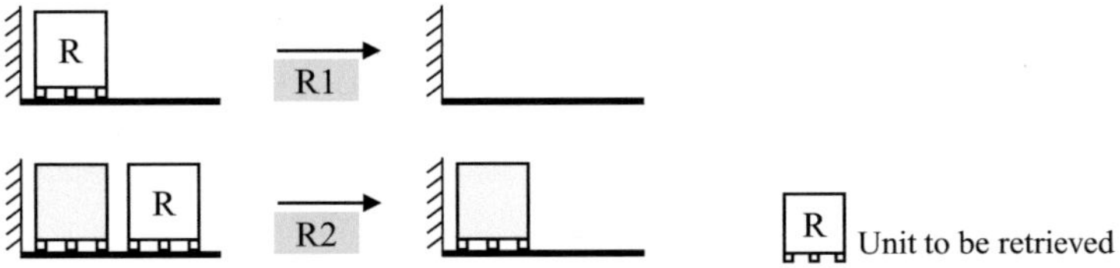

Figure 4.5: Retrieval events without rearrangement

The probabilities, that a retrieval operation is performed as event R1 or R2, are:

$$P(R1) = \frac{P(H)}{P(H) + 2P(F)} \tag{4.8}$$

$$P(R2) = \frac{P(F)}{P(H) + 2P(F)} \tag{4.9}$$

Units for retrieval can be chosen from half-filled storage lanes (H) or occupied storage lanes (F). For those storage lanes in state F, there is the possibility to either chose the unit at the front or at the rear position for retrieval. Consequently, occupied storage lanes have twice the chance of being chosen and P(F) is doubled in the denominator. R2 denotes the case of choosing the unit at the front.

Example 4.1 *To understand the denominator of equations 4.8 and 4.9 consider the following: In a storage space, there are five fully occupied storage lanes (state F) and five half-filled storage lanes (state H). In total, 15 units are stored and* $P(H) = 1/2$ *and* $P(F) = 1/2$*. Each unit has the same probability of selection.*

Determine the probability of choosing a unit at the rear of a filled storage lane for retrieval: Intuitively, the probability is $5/15 = 1/3$*. The probability that a randomly selected, non-empty storage lane is completely filled is* $1/2$*. For a half-filled storage lane, the probability is* $1/2$ *too. Based on this, choosing a unit at the rear from a filled storage lane is* $\frac{1/2}{1/2+2\cdot 1/2} = 1/3$*.*

2P(F) in the denominator allows that both positions of a filled storage lane are taken into account.

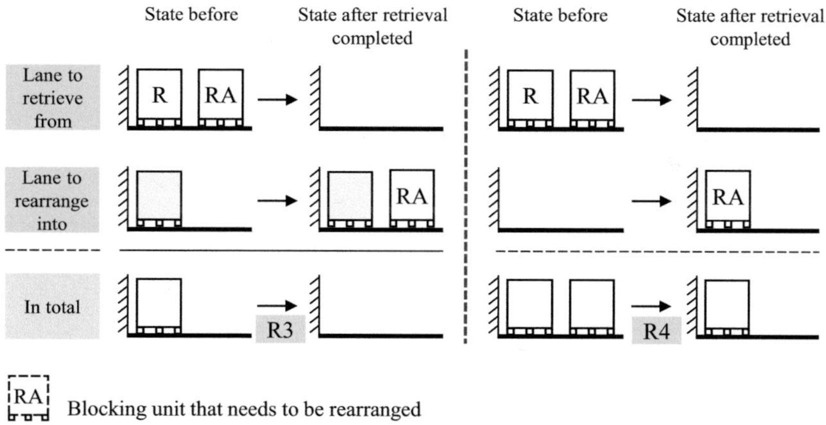

Figure 4.6: Rearranging into an half-filled lane (left) and into an empty storage lane (right)

Next, we consider the retrieval operations that cause a **rearrangement**. Two storage lanes are affected by retrieval operations with regular rearrangements: The storage lane to retrieve from and the the storage lane to re-store the blocking unit. The storage lane that is retrieved from becomes an empty storage lane. The storage lane used for the rearrangement is either empty or half-filled. Therefore, re-storing during rearrangements follows the same logic than storing and we distinguish two events (see S1 and S2). For both events, we observe a combined state transition *in total* that is the consequence of the two affected storage lanes (see Figure 4.6).

R3 describes the case in which a half-filled storage lane is used for rearrangement. In total, with a retrieval of type R3, a half-filled storage lane disappears while an empty storage lane emerges. **R4** describes the situation that an empty storage lane is chosen for rearranging the blocking unit into. In total, a transition from a filled storage lane to a half-filled storage lane is observed.

R3 and R4 are shown in Figure 4.6. The first row shows the lane the unit is retrieved from, while the second row shows the storage lane used for rearrangement. At the bottom, in each case the total transition is shown.

R5 describes the retrieval of a unit using the tango. In this case, a fully occupied storage lane transfers into a half-filled storage lane as the blocking unit from the front position is restored to the rear position of the same storage lane. The transition is shown in Figure 4.7.

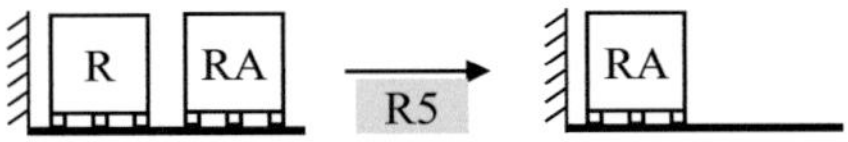

Figure 4.7: Retrieval event when performing tango

To determine the probabilities of R3, R4 and R5, the probability distribution of the **execution order** is required. Regular rearrangements, i.e., R3 and R4, occur in both execution orders, while tango, R5, occurs only in the SSRR order. For a random retrieval operation within a cycle, it is equally likely being the first or second retrieval operation. As rearrangements at the first retrieval within the SSRR-path are always performed as a tango, there is a probability of $1/2$ that a rearrangement is performed as a tango within SSRR. Within SRSR, all rearrangements are performed in the regular way.

Therefore, the probability of transition R3 consists of three elements:

- The probability of choosing the rear unit of a filled storage lane for retrieval.
- The probability of rearranging into a storage lane in state H.
- The conditional probability of a regular rearrangement given a rearrangement takes place $(1 - 1/2P(SSRR))$.

In the same way, R4 is composed.

$$P(R3) = \frac{P(F)}{P(H)+2P(F)} \cdot \frac{P(H)}{P(E)+P(H)} \cdot (1 - \tfrac{1}{2}P(SSRR)) \tag{4.10}$$

$$P(R4) = \frac{P(F)}{P(H)+2P(F)} \cdot \frac{P(E)}{P(E)+P(H)} \cdot (1 - \tfrac{1}{2}P(SSRR)) \tag{4.11}$$

The probability of R5 is composed of the probability of choosing the rear unit of a filled storage lane for retrieval and the conditional probability for a tango rearrangement $(1/2P(SSRR))$:

$$P(R5) = \frac{P(F)}{P(H)+2P(F)} \cdot \tfrac{1}{2}P(SSRR) \tag{4.12}$$

At the same time, equation 4.12 gives the unconditional probability of performing a tango, $P(Tango)$.

The rearrangement probability is composed of these three transitions.

$$P_{Rearrange} = P(R3) + P(R4) + P(R5) = \frac{P(F)}{P(H)+2P(F)} \tag{4.13}$$

The probability of performing a regular rearrangement is composed of the probabilities for the events R3 and R4. Therefore, the unconditional probability of performing a regular rearrangement can be written as follows:

$$\begin{aligned} P(Regular) &= P(R3) + P(R4) \\ &= \frac{P(F)}{P(H)+2P(F)} \cdot (1 - \tfrac{1}{2}P(SSRR)) \end{aligned} \tag{4.14}$$

The unconditional probability of performing a tango rearrangement is defined by

$$P(Tango) = P(R5) = \frac{P(F)}{P(H)+2P(F)} \cdot \frac{1}{2}P(SSRR) \tag{4.15}$$

Summarizing all events All possible state transitions of a storage lane are summarized in Figure 4.8.

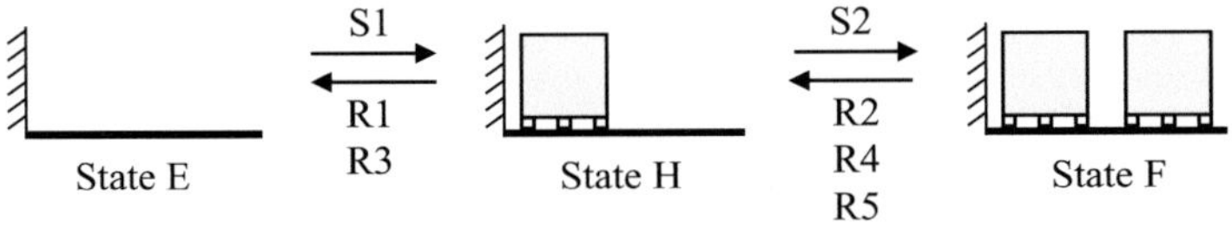

Figure 4.8: All possible state transitions of a storage lane

Similarly to the conditional probabilities of choosing a specific type of a storage lane for storage operations (see equation 4.6 and 4.7), it is possible to define these probabilities for retrieval operations. We can distinguish between the possibility of choosing either a filled or a half-filled storage lane for retrieval. The probability to choose a half-filled storage lane, given a retrieval takes place, is determined by the sum of R1 and R3 (see Figure 4.8).

$$\begin{aligned} P(H|R) &= P(R1) + P(R3) \\ &= \frac{P(H)}{P(H)+2P(F)} \cdot \left(1 + \frac{P(F)}{P(E)+P(H)} \cdot (1 - \tfrac{1}{2}P(SSRR))\right) \end{aligned} \tag{4.16}$$

Accordingly, the probability to choose a fully occupied storage lane for retrieval is given by the sum of R2, R4 and R5.

$$\begin{aligned} P(F|R) &= P(R2) + P(R4) + P(R5) \\ &= \frac{P(F)}{P(H)+2P(F)} \cdot \left(1 + \frac{P(E)}{P(E)+P(H)} \cdot (1 - \tfrac{1}{2}P(SSRR)) + \tfrac{1}{2}P(SSRR)\right) \end{aligned} \tag{4.17}$$

Modeling as a Markov Chain

To obtain the probability distribution of the storage lanes' states (E, H or F) and subsequently being able to compute equation 4.13, the allocation of the storage lanes is modeled as a stochastic process (Lippolt 2003). A storage lane can only take one single state at a specific point of time. State transitions are caused by storage or retrieval operations. This stochastic process has a finite state space, which is $S = \{E, H, F\}$. Transitions only occur at discrete points of time and transition probabilities only depend on the current state of a storage lane. Therefore, this process can be described as a time discrete Markov Chain. As shown in Figure 4.8, every state can be reached from any other state, which means the Markov Chain is irreducible. Moreover, the chain has an aperiodic state with $p_{ii} > 0, i \in S$, as in every step all storage lanes do a transition, but most of them do a transition to the same state. An illustration of the Markov Chain is shown in Figure 4.9. For irreducible, aperiodic Markov Chains, the stationary distribution π can be described by the asymptotic behavior for $t \to \infty$. π is the unique solution of the system of equations 3.11 - 3.13 and gives the probability distribution of the storage lane allocation in a steady state.

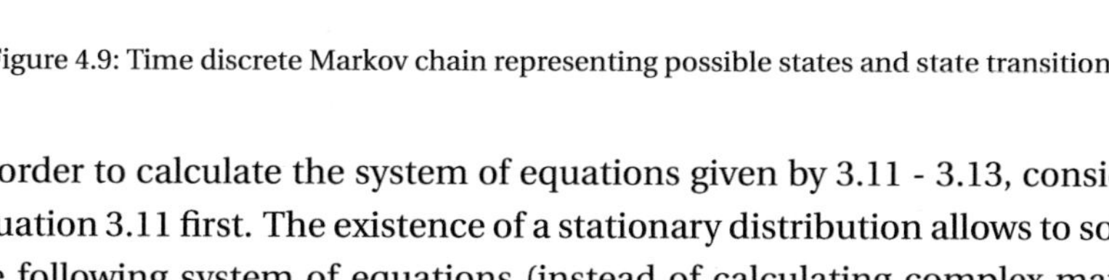

Figure 4.9: Time discrete Markov chain representing possible states and state transitions

In order to calculate the system of equations given by 3.11 - 3.13, consider equation 3.11 first. The existence of a stationary distribution allows to solve the following system of equations (instead of calculating complex matrix operations):

$$\pi = \pi P \tag{4.18}$$

The transition matrix is P, while the stationary distribution π is its unique solution.

Transition matrix The transition probabilities represent the likelihood of a storage or retrieval event, given a certain state of the storage lane (represented as edge weights in Figure 4.9). Note that the possibility of staying in a certain state also represents a relevant transition.

It is important to point out that the probabilities defined in the last section do not represent the required transition probabilities. Instead, they give the probability that a storage lane is affected (or say chosen), given that a certain operation takes place. However, we can use these probabilities to formulate the transition probabilities when applying Bayes' theorem for conditional probabilities (Bayes and Price 1763).

Consider the conditional probability that a storage operation affects a half filled lane, which is $P(H \mid S)$ (see equation 4.7). For the Markov Chain, the probability that, given the storage lane is in state H, a storage event happens ($P(S \mid H)$), is required. By applying the Bayes' theorem we can derive:

$$\begin{aligned} P(S \mid H) &= \frac{P(H|S) \cdot P(S)}{P(H)} \overset{P(H|S)=P(S2)}{=} \frac{P(S2) \cdot P(S)}{P(H)} \\ &= \frac{P(H) \cdot P(S)}{(P(E)+P(H)) \cdot P(H)} = \frac{P(S)}{P(E)+P(H)} \end{aligned} \tag{4.19}$$

The other state transitions are calculated analogously:

$$P(S \mid E) = \frac{P(S)}{P(E)+P(H)} \tag{4.20}$$

$$P(R \mid H) = \frac{P(R) \cdot (P(E)+P(H)+P(F)(1-\frac{1}{2}P(SSRR)))}{(P(H)+2P(F))(P(E)+P(H))} \tag{4.21}$$

$$P(R \mid F) = \frac{P(R) \cdot (2P(E)+P(H)(1+\frac{1}{2}P(SSRR)))}{(P(H)+2P(F))(P(E)+P(H))} \tag{4.22}$$

The transition matrix P shows all possible state transitions.

$$P = \begin{array}{c} \\ E \\ H \\ F \end{array} \begin{array}{c} \begin{array}{ccc} E & H & F \end{array} \\ \begin{bmatrix} 1-P(S\mid E) & P(S\mid E) & 0 \\ P(R\mid H) & 1-P(R\mid H)-P(S\mid H) & P(S\mid H) \\ 0 & P(R\mid F) & 1-P(R\mid F) \end{bmatrix} \end{array} \tag{4.23}$$

Stationary storage lane allocation We can set up the following system of linear equations according to equation 4.18:

$$\begin{aligned} P(E) &= (1-P(S|E))P(E) + P(R|H)P(H) + 0P(F) \\ P(H) &= P(S|E)P(E) + (1-P(R\mid H) - P(S\mid H))P(H) + P(R|F)P(F) \\ P(F) &= 0P(E) + P(S|H)P(H) + (1-P(R\mid F))P(F) \end{aligned} \tag{4.24}$$

Because of its linear dependence, system 4.24 can be reduced to a single equation and we only consider the first equation of 4.24, which can be seen as equilibrium condition for state E.

$$P(E)P(S|E) = P(H)P(R|H) \tag{4.25}$$

Using 4.20 and 4.21 yields:

$$\frac{P(F)\,P(S)}{P(E)+P(H)} = \frac{P(H)\,P(R)\,(P(E)+P(H)+P(F)(1-\frac{1}{2}P(SSRR)))}{(P(H)+2P(F))(P(E)+P(H))} \tag{4.26}$$

This expression contains the probabilities of storage and retrieval in general, $P(S)$ and $P(R)$. While only complete quadruple command cycles are conducted, storage and retrieval operations are balanced on average and $P(S)$ and $P(R)$ can be canceled. We obtain the probability that a storage lane is in state F, depending on the two others states:

$$P(F) \overset{P(S)=P(R)}{=} \frac{P(H)^2}{2P(E) - P(H)(1-\frac{1}{2}P(SSRR))} \tag{4.27}$$

In addition to 4.27 (which is derived from 3.11), we incorporate 3.12 and 3.13 to the system of equations and add an additional equation to include the filling level z as parameter. This leads to a solvable system of equations that is depending on the filling level. Using equation 4.2, we obtain the following system of equations:

$$\begin{aligned} 1 &= P(E) + P(H) + P(F) \\ z &= \frac{2 \cdot P(F) + P(H)}{2} \\ P(F) &= \frac{P(H)^2}{2P(E) - P(H)(1 - \frac{1}{2}P(SSRR))} \\ P(E), P(H), P(F) &\geq 0 \end{aligned} \tag{4.28}$$

When solving this system, we obtain two possible solutions due to the squared expression in the equilibrium condition. According to the last equation of 4.28, only positive solutions are valid. Moreover, both domain and co-domain are defined by the interval $[0,1]$, as the probabilities are expressed in dependence of the filling level. For each probability, $P(E), P(H)$ and $P(F)$, one possible solutions lies outside the defined interval and is rejected. We obtain the following results for the state probabilities of the storage lanes, depending on the filling level, z, and $P(SSRR)$.

$$P(E) = \frac{2z + P(SSRR)z - \sqrt{P(SSRR)^2 z^2 - 12P(SSRR)z^2 + 4P(SSRR)z + 4z^2 + 8z + 4} + 2}{2P(SSRR)} - 2z + 1 \tag{4.29}$$

$$P(H) = 2z - \frac{2z + P(SSRR)z - \sqrt{P(SSRR)^2 z^2 - 12P(SSRR)z^2 + 4P(SSRR)z + 4z^2 + 8z + 4} + 2}{P(SSRR)} \tag{4.30}$$

$$P(F) = \frac{2z + P(SSRR)z - \sqrt{P(SSRR)^2 z^2 - 12P(SSRR)z^2 + 4P(SSRR)z + 4z^2 + 8z + 4} + 2}{2P(SSRR)} \tag{4.31}$$

$P(E), P(H)$ and $P(F)$ represent the the stationary distribution. The expressions 4.29 to 4.31 are not valid for $P(SSRR) = 0$. We present a limit evaluation in the Appendix A and show that all terms converge to the probabilities presented by Lippolt (2003).

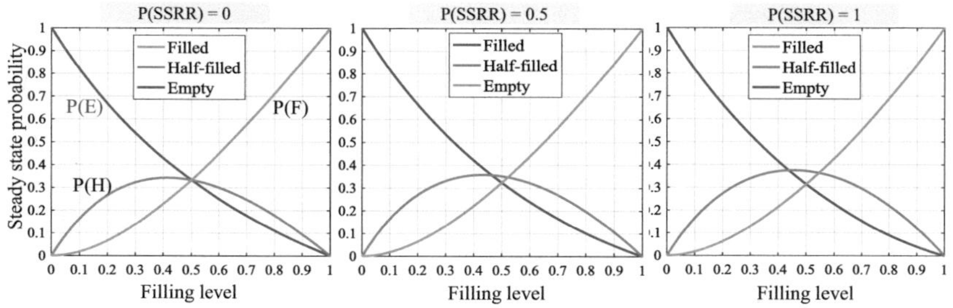

Figure 4.10: Steady state probabilities for different values of $P(SSRR)$

Figure 4.10 shows the state probabilities of the storage lanes, depending on the filling level, for three different values of $P(SSRR)$. We see that the curve representing storage lanes in state H is skewed upwards with increasing $P(SSRR)$. This is accredited to an increased ratio of half-filled storage lanes caused by a higher amount of tango rearrangements. With an increasing amount of tango (increasing $P(SSRR)$), the number of half filled lanes increases. Therefore, to correctly define the mean storage lane allocation, the amount of tango in the cycle has to be taken into account. Figure 4.11 illustrates this behavior of half-filled storage lanes. All curves of Figure 4.10 are consolidated in one diagram (Figure 4.11, left-hand side). Additionally, a 3D surface plot shows $P(H)$ depending on P(SSRR) and the filling level (right-hand side).

From equation 4.13, we derive the probabilities of performing a rearrangement using the results of the equations 4.29 to 4.31:

$$P_{Rearrange} = \frac{2\,z + \mathrm{P(SSRR)}\,z - \sqrt{\mathrm{P(SSRR)}^2\,z^2 - 12\,\mathrm{P(SSRR)}\,z^2 + 4\,\mathrm{P(SSRR)}\,z + 4\,z^2 + 8\,z + 4} + 2}{4\,\mathrm{P(SSRR)}\,z} \tag{4.32}$$

With equations 4.14 and 4.15, the rearrangement probability for both regular and tango rearrangement can be specified.

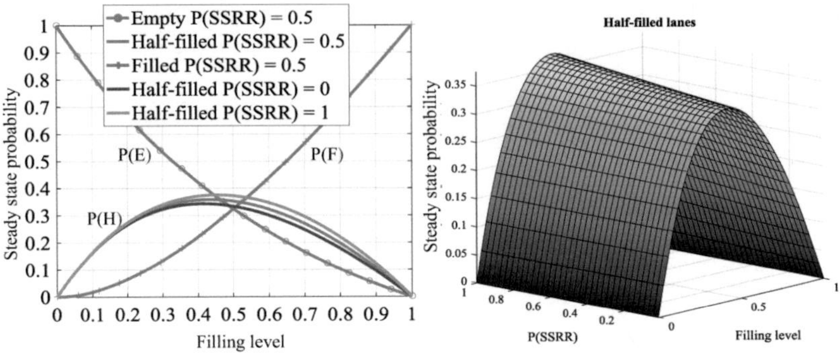

Figure 4.11: Illustration of different steady state probabilities for state H for different P(SSRR)

4.2.2 Rearrangement- and Load Handling- Cycle Times

The remaining components of the travel time formula depend on the state probabilities of the storage lanes. In this subsection we address the **mean cycle times** of **regular** and **tango rearrangements**. Subsequently, the **mean handling times** of the LHD are considered.

Regular rearrangement Time

In order to determine the expected rearrangement cycle time, $E(RC)$, the position of the nearest available storage lane for rearrangement is needed. Apparently, this depends on the filling level, or the number of storage locations in state E or H. At a low filling level, there is a relatively high probability of neighboring storage lanes being either empty or half-filled. With an increasing filling level, this probability decreases, hence the nearest available position moves further away. The mean distance to the nearest rearrangement position is represented by the nearest of m randomly distributed, open locations. Han et al. (1987) analytically determine the smallest of m random distances under the assumptions of a continuous rack model formulated that needs to be solved numerically (see equation 3.59). Lippolt

(2003) derives a closed form approximation for the distance of the nearest rearrangement position in a discrete rack model (see subsection 3.3.2).

According to Lippolt (2003, p. 154), we define the expected distance to the nearest rearrangement position , $E(RD)$, by:

$$E(RD) = \left(\frac{7}{15}\right)^{1-\frac{pR}{l}} \cdot \frac{1}{\sqrt{pR}} \tag{4.33}$$

pR denotes the number of potential storage lanes, available for rearrangement and is calculated by multiplying the probability that a storage lane is qualified as rearrangement location (i.e., state E or H) with the total number of storage lanes (l). With the state probabilities provided above (equations 4.29, 4.30 and 4.31, pR results in:

$$pR = (P(E) + P(H)) \cdot l \tag{4.34}$$

Equation 4.33 denotes the expected path-depending cycle time of a rearrangement command $E(RC)$ in the normalized rack.

Tango Time

The tango rearrangement is a **predefined movement** of the S/R machine and is free from stochastic effects. Consequently, it provides a **deterministic handling time** that is only depending on the characteristics of the S/R machine. The tango consists of four steps:

1. Picking from the front position.
2. Moving the S/R machine sideways to position the other load handling device for picking from the same storage lane.
3. Picking the retrieval unit from the rear position.
4. Moving back and storing the unit retrieved first back to the rear position.

Therefore, the path-depending time needed to perform a tango can be calculated in the following way:

$$t_{LHD,tango} = t_{LHD,f} + t_{LHD,r} + 2 \cdot t_d \tag{4.35}$$

with t_d being the time needed to move sideways in order to position the other load handling device. d represents the distance between the LHDs and, at the same time, the distance the load handling device needs to move. Note that the access time to the rear position is incorporated only once in equation 4.35, because accessing and handling of the retrieval unit is not part of the actual rearrangement.

Mean Load Handling Times

Additionally, access times of the LHD need to be incorporated into the expression. The load handling times given by $t_{LHD,f}$ and $t_{LHD,r}$ describe **one access cycle** with extending (into the rack) and retracting of the device. For one type of operation, two access cycles arise: The **pick-up** and the **deposit** of the storage unit. For storage operations, the pick-up always takes place at the I/O point, while for retrieval operations it is performed at the storage position inside the rack. The other way round, during storage operations the deposition takes place inside the rack, while retrieval operations are deposited at the I/O point.

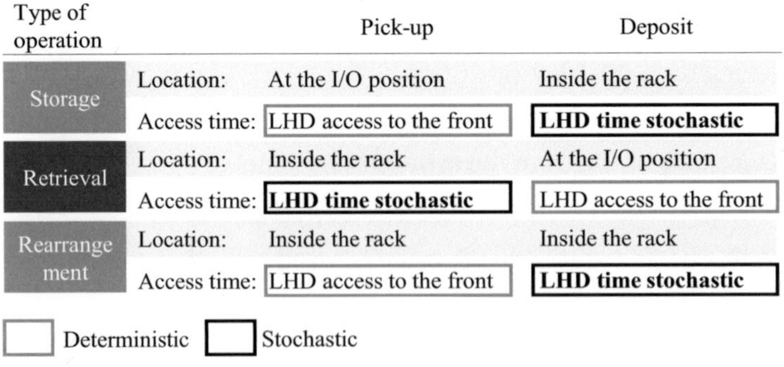

Figure 4.12: Distribution of access times for both storage and retrieval operations

Figure 4.12 gives an overview of the different access times of the LHD that occur for storage, retrieval and rearrangements. Note that for storage and rearrangement operations, the handling times for pick-up and deposit are equal. Therefore, we do not considered them separately.

The **pick-up times** for **storage** and **rearrangement** are **deterministic** and given by the time to access the front position ($t_{LHD,f}$). For storage operations pick-up occurs is at the I/O point, while units that are rearranged, are always picked-up from a front position due to their blocking of another unit. The **deposition** time for **retrieval** operations is known in advance and therefore **deterministic**. Retrieval units are deposited at the I/O point, which requires the time to access the front position. As a result, every load handling cycle with double deep storage has one deterministic and one stochastic part.

The **stochastic** part depends on the position of the storage lane that is affected, i.e., the front or the rear position. The state of the storage lane is depending on the filling level.

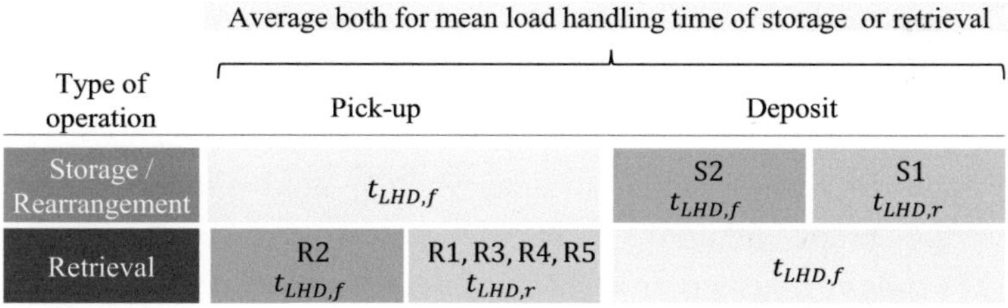

Figure 4.13: Composition of the load handling times for storage and retrieval operations

For the derivation of the stochastic part, the the different types of storage and retrieval events are considered. For storage operations, the probabilities of storing into the front position (S2) and of storing into the rear position (S1) are weighted with the access times for front position ($t_{LHD,f}$) and rear position ($t_{LHD,r}$), respectively (see the first row of Figure 4.13). Load handling times for rearrangements are equivalent, as the same options are available for the re-storage part of rearrangements.

To obtain the mean load handling time for one type of operation, we average the deterministic and the stochastic part of each type of operation. The mean load handling time of a storage operation, which represents the average of picking-up at the I/O position and deposition at the storage position, results in:

$$E(t_{LHD}^{S}) = \frac{1}{2} \cdot t_{LHD,f} + \frac{1}{2} \cdot \left(t_{LHD,f} \cdot P(S2) + t_{LHD,r} \cdot P(S1)\right) \tag{4.36}$$

For retrieval operations, we follow the same approach and the mean load handling times can be specified by:

$$E(t_{LHD}^{S}) = \frac{1}{2} \cdot \left(t_{LHD,f} \cdot (1 + P(S2)) + t_{LHD,r} \cdot (1 - P(S2))\right) \tag{4.37}$$

$$E(t_{LHD}^{R}) = \frac{1}{2} \cdot \left(t_{LHD,f} \cdot (1 + P(R2)) + t_{LHD,r} \cdot (1 - P(R2))\right) \tag{4.38}$$

The presented mean load handling times are not affected by the fact that we use two handling devices.

4.2.3 Composing the Travel Time Formula

We assume the travel time for a QC to be composed of the travel time of the QC in single deep storage environments ($E(QC)$) plus expected travel times for rearrangement operations. The derivation of each component is discussed in the previous sections. The eventual composition of all elements is the objective of this subsection. Therefore, equation 4.5 is completed by adding mean load handling times, dwell times and denormalization.

Load handling

Mean load handling times for storage, rearrangement and retrieval operations need to be added to the formula. Consider the following two aspects:

1. For each operation (storage, rearrangement or retrieval), the particular mean load handling time is added twice, allowing for pick-up

and deposit. This means for the two storage operations the term is $2 \cdot 2 \cdot E(t^S_{LHD})$.

2. The assumptions for the handling of the units at the I/O point are important. We assume that both storage units are each picked-up and deposited at the same time. For such simultaneous handling of the units, the access time at the I/O point is subtracted twice, i.e. $2 \cdot t_{LHD,f}$.

Consequently, the addition to the formula is $4 \cdot E(t^S_{LHD}) + 4 \cdot E(t^R_{LHD}) - 2 \cdot t_{LHD,f}$ for the part referring to the storage and retrieval operations and $2 \cdot E(t^S_{LHD})$ for the rearrangement command.

Additional times and denormalization

Moreover, we need to rescale the formula by $\frac{L}{v_x} = T$. To obtain the real time needed to perform a cycle, dwell times and both acceleration and deceleration need to be taken into account. Typically, there is one general addition for dwell times often referred to as t_0. For each movement of the AS/RS machine, the term $\frac{1}{2} \cdot \left(\frac{v_x}{a_x} + \frac{v_y}{a_y}\right)$ is added due to acceleration and deceleration. Considering these aspects, the de-normalized time for the quadruple command cycle is:

$$\begin{aligned} E(QC)_{dd} = & t_0 + \frac{5}{2} \cdot (\frac{v_x}{a_x} + \frac{v_y}{a_y}) + E(QC)^N \cdot \frac{L}{v_x} \\ & + P(SSRR) P_{Rearrange} \cdot t_{Tango} \\ & + (1 + (1 - P(SSRR)) P_{Rearrange} \cdot E(RC) \\ & + 4 \cdot E(t^S_{LHD}) + 4 \cdot E(t^R_{LHD}) - 2 \cdot t_{LHD,f} \end{aligned} \tag{4.39}$$

and

$$t_0 = 12 \cdot t_{dead} + 5 \cdot t_{mast} \tag{4.40}$$

$$t_{Tango} = t_{LHD,tango} + 4 \cdot t_{dead} + 2 \cdot t_{mast} \tag{4.41}$$

The rearrangement command is composed accordingly:

$$E(RC) = t_{0,Rearr.} + 2 \cdot E(t^S_{LHD}) + (\frac{v_x}{a_x} + \frac{v_y}{a_y}) + 2 \cdot E(RD) \cdot \frac{L}{v_x} \quad (4.42)$$

with

$$t_{0,Rearr} = 4 \cdot t_{dead} + 2 \cdot t_{mast} \quad (4.43)$$

The expected rearrangement distance ($E(RD)$) is determined according to equation 4.33.

For a quadruple command cycle, we assume 12 occurrences of dead times, which is two per access cycle of the LHD (i.e., pick-up at the I/O point, four stops within the cycle and deposition at the I/O point). For every movement of the S/R machine or travel between positions, one t_{mast} is assumed. Note that depending on the technical situation, $2 \cdot t_{mast}$ may be excluded from equation 4.41. If the mast needs no damping after small and slow movements during tango, or if the tango can be performed without moving the mast, this could be possible.

4.3 Calculation Examples for Quadruple Command Cycles

In this section, we present the results for specific values of the execution order's probability distribution. The examples represent the two extreme cases of $P(SSRR;SRSR) = (0,1)$ and $P(SSRR;SRSR) = (1,0)$ as well as a case where both execution orders are randomly selected ($P(SSRR;SRSR) = (1/2, 1/2)$ (see Table 4.1).

The steady state probabilities of the storage lanes have impact on the following aspects:

- Rearrangement probability,
- Number of potential storage lanes for performing a rearrangement and the mean rearrangement distance, respectively,
- Mean load handling times.

Consequently, for different values of $P(SSRR)$, we obtain different variants of the mean cycle time for a quadruple command cycle.

4.3.1 Basic Model

The basic model represents a cycle with a low level of control. Both execution orders are possible and have an equal probability to occur ($P(SSRR) = P(SRSR) = 1/2$). We obtain the following results:

- Stationary storage lane allocation

$$P(E) = -\frac{1}{2}\sqrt{-7\cdot z^2+40\cdot z+16}+\frac{1}{2}\cdot z+3 \tag{4.44}$$

$$P(H) = \sqrt{-7\cdot z^2+40\cdot z+16}-3\cdot z-4 \tag{4.45}$$

$$P(F) = -\frac{1}{2}\sqrt{-7\cdot z^2+40\cdot z+16}+\frac{5}{2}\cdot z+2 \tag{4.46}$$

- Rearrangement probability

$$P_{Rearrange} = \frac{5\,z-\sqrt{-7\,z^2+40\,z+16}+4}{4\,z} \tag{4.47}$$

- Number of potential storage lanes for rearrangement

$$pR = (\frac{1}{2}\sqrt{-7\,z^2+40\,z+16}-\frac{5\,z}{2}-1)\cdot l \tag{4.48}$$

- Mean load handling times

When inserting the probabilities for the events of storage and retrieval, we obtain the load handling times as a function of the filling level:

$$
\begin{aligned}
E(t^S_{LHD}) =& \frac{1}{2} \cdot (t_{LHD,f} \cdot (1 + \frac{-3z + \sqrt{-7 \cdot z^2 + 40 \cdot z + 16} - 4}{-\frac{5}{2}z + \frac{1}{2}\sqrt{-7 \cdot z^2 + 40 \cdot z + 16} - 1}) \\
&+ t_{LHD,r} \cdot (1 - \frac{-3z + \sqrt{\cdots} - 4}{-\frac{5}{2}z + \frac{1}{2}\sqrt{\cdots} - 1}))
\end{aligned}
\tag{4.49}
$$

$$
\begin{aligned}
E(t^R_{LHD}) =& \frac{1}{2} \cdot (t_{LHD,f} \cdot \frac{9z - \sqrt{-7 \cdot z^2 + 40 \cdot z + 16} + 4}{4z}) \\
&+ \frac{1}{2} \cdot (t_{LHD,r} \cdot \frac{-z + \sqrt{-7 \cdot z^2 + 40 \cdot z + 16} - 4}{4z})
\end{aligned}
\tag{4.50}
$$

Mean travel time

The expected cycle time of a quadruple command cycle with an equal probability for both execution orders is:

$$
\begin{aligned}
E(QC_{dd}) =& t_0 + \frac{5}{2} \cdot (\frac{v_x}{a_x} + \frac{v_y}{a_y}) + E(QC)^N \cdot \frac{L}{v_x} \\
&+ \frac{3}{2} \cdot P_{Rearrange} \cdot E(RC) + \frac{1}{2} \cdot P_{Rearrange} \cdot t_{Tango} \\
&+ 4 \cdot E(t^S_{LHD}) + 4 \cdot E(t^R_{LHD}) - 2 \cdot t_{LHD,f}
\end{aligned}
\tag{4.51}
$$

$E(RC)$ is adjusted according to equations 4.42 and 4.33.

4.3.2 SSRR Model

The execution order is limited to the case SSRR. This order is closer linked to the traditional order of performing a dual command cycle ('Storage-Retrieval') and also considered by Xu, Shen, Yugang Yu and Huang (2015).

By applying SSRR only ($P(SSRR) = 1$ and $P(SRSR) = 0$), tango rearrangements occur more frequently. The results are:

- Stationary storage lane allocation

$$P(E) = 2 - \frac{\sqrt{-7 \cdot z^2 + 12 \cdot z + 4}}{2} - \frac{z}{2} \tag{4.52}$$

$$P(H) = \sqrt{-7 \cdot z^2 + 12 \cdot z + 4} - z - 2 \tag{4.53}$$

$$P(F) = \frac{3z}{2} - \frac{\sqrt{-7 \cdot z^2 + 12 \cdot z + 4}}{2} + 1 \tag{4.54}$$

- Rearrangement probability

$$P_{Rearrange} = \frac{3z - \sqrt{-7 \cdot z^2 + 12 \cdot z + 4} + 2}{4z} \tag{4.55}$$

- Number of potential storage lanes for rearrangement

$$pR = (\frac{1}{2}\sqrt{-7 \cdot z^2 + 12 \cdot z + 4} - \frac{3}{2}z) \cdot l \tag{4.56}$$

- Mean load handling times

$$\begin{aligned} E(t^S_{LHD}) = & \frac{1}{2} \cdot (t_{LHD,f} \cdot (1 + \frac{z - \sqrt{-7 \cdot z^2 + 12 \cdot z + 4} + 2}{\frac{3}{2}z - \frac{1}{2}\sqrt{-7 \cdot z^2 + 12 \cdot z + 4}}) \\ & + t_{LHD,r} \cdot (1 - \frac{z - \sqrt{-7 \cdot z^2 + 12 \cdot z + 4} + 2}{\frac{3}{2}z - \frac{1}{2}\sqrt{-7 \cdot z^2 + 12 \cdot z + 4}})) \end{aligned} \tag{4.57}$$

$$\begin{aligned} E(t^R_{LHD}) = & \frac{1}{2} \cdot (t_{LHD,f} \cdot (1 + \frac{\frac{3}{2}z - \frac{1}{2}\sqrt{-7 \cdot z^2 + 12 \cdot z + 4} + 1}{2z}) \\ & + t_{LHD,r} \cdot (1 - \frac{\frac{3}{2}z - \sqrt{-7 \cdot z^2 + 12 \cdot z + 4} + 1}{2z})) \end{aligned} \tag{4.58}$$

Mean travel time

The expected cycle time of a quadruple command cycle following a SSRR execution order is:

$$
\begin{aligned}
E(QC_{dd}) = t_0 + \frac{5}{2} \cdot (\frac{v_x}{a_x} + \frac{v_y}{a_y}) + E(QC)^N \cdot \frac{L}{v_x} \\
+ 1 \cdot P_{Rearrange} \cdot E(RC) + 1 \cdot P_{Rearrange} \cdot t_{Tango} \\
+ 4 \cdot E(t_{LHD}^S) + 4 \cdot E(t_{LHD}^R) - 2 \cdot t_{LHD,f}
\end{aligned}
\tag{4.59}
$$

$E(RC)$ is adjusted according to equations 4.42 and 4.33.

4.3.3 SRSR Model

Performing a QC without a tango, might have two different reasons: Either the execution sequence is limited to SRSR, which does not allow tango operations. Or there are technical circumstances that prevent the realization of tango, e.g. the controlling computer does not support tango. Independent of the real execution order, the second case can be modeled in that way where $P(SSRR) = 0$ and $P(SRSR) = 1$ apply. The results of this case are equivalent to Lippolt. All expressions can be found in (Lippolt 2003, p. 135):

- Stationary storage lane allocation

$$P(E) = \frac{1-z}{1+z} \tag{4.60}$$

$$P(H) = \frac{2z(1+z)}{1+z} \tag{4.61}$$

$$P(F) = \frac{2z^2}{1+z} \tag{4.62}$$

- Rearrangement probability

$$P_{Rearrange} = \frac{z}{1+z} \tag{4.63}$$

Note that $P(Tango) = 0$. Since the quadruple command cycle consists of two retrievals, the rearrangement probability presented by Lippolt is doubled.

- Number of potential storage lanes for rearrangement

$$pR = \frac{(1+z) - 2z^2}{1+z} \cdot l \tag{4.64}$$

- Mean load handling times

$$E(t_{LHD}^S) = \frac{t_{LHD,f} \cdot (1+4z) + t_{LHD,r}}{2(1+2z)} \tag{4.65}$$

$$E(t_{LHD}^R) = \frac{t_{LHD,f} \cdot (1+2z) + t_{LHD,r}}{2(1+z)} \tag{4.66}$$

Mean travel time

Using the terms stated above in the following equation (4.67), we obtain the cycle time for a quadruple command cycle without tango.

$$\begin{aligned} E(QC_{dd}) = {} & t_0 + \frac{5}{2} \cdot (\frac{v_x}{a_x} + \frac{v_y}{a_y}) + E(QC)^N \cdot \frac{L}{v_x} \\ & + 2 \cdot P_{Rearrange} \cdot E(RC) \\ & + 4 \cdot E(t_{LHD}^S) + 4 \cdot E(t_{LHD}^R) - 2 \cdot t_{LHD,f} \end{aligned} \tag{4.67}$$

$E(RC)$ is adjusted according to equations 4.42 and 4.33.

Equation 4.67 is similar to the model for a quadruple command cycle with double deep storage under random storage policy presented by Seemüller (2006). Both are derived by applying the probabilities of Lippolt (2003) because of the *no Tango* assumption. In contrast to Seemüller (2006), we use the rearrangement travel time proposed by Lippolt (2003), whereas they use a simplified approximation.

4.4 Model of a Quadruple Command Cycle with Modified Tango

In this section we present a modified model with a more flexible execution of the tango movement by changing the assumptions of section 4.1.1. So far, we strictly assume that a tango rearrangement is performed according to the procedure shown in Figure 4.2. However, the time to perform a tango rearrangement can be shortened as shown in Figure 4.14. Consider a tango rearrangement while one of the storage lanes next to the retrieval position is either in state E or H. In this case, the blocking unit may be restored into the respective lane. Consequently, the load handling device does not need to move back to restore the unit to the rear position of the original storage lane (see Figure 4.14, pictures c and d).

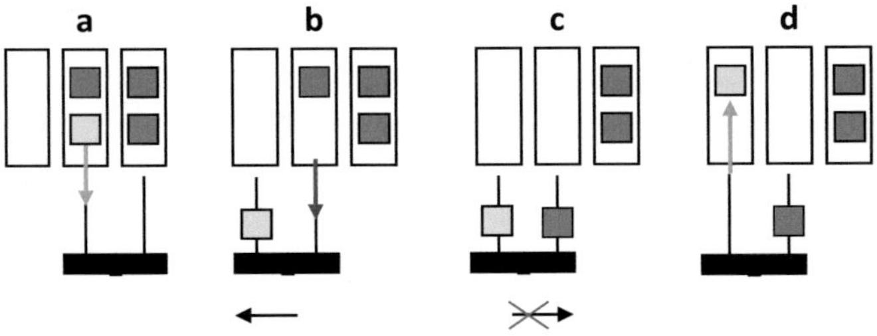

Figure 4.14: Possibility to shorten tango movement

The blocking unit is always re-stored to the rear position during the execution of a standard tango. When executing the modified tango as explained above, there are **two possible scenarios** for the storage lane in which the re-storage takes place:

1. A neighboring storage lane is empty (E) and the blocking unit is re-stored to the rear position, as shown in Figure 4.14. In this way, the time needed to move the LHDs sideways from one lane to a neighboring lane, t_d, is saved.

2. A neighboring storage lane is half-filled (H) and the blocking unit is re-stored to the front position. In this way, the saving is t_d plus the difference between the time needed to access the rear or the front position, i.e., $t_{LHD,r} - t_{LHD,f}$.

The savings of the modified tango are depending on the state of the neighboring storage lanes and are specified in Table 4.2. As the savings are positive in both cases, the modified tango is always beneficial. Consequently, as soon as one of the adjacent storage lanes provides at least one free position, the modified tango is performed.

State of the adjacent storage lane	LHD time for re-storage	Savings
E	$t_{LHD,r}$	t_d
H	$t_{LHD,f}$	$t_d + (t_{LHD,r} - t_{LHD,f})$

Table 4.2: Possible savings of performing a modified tango depending on the state of the adjacent lanes compared to performing a standard tango

For an analytical representation of a QC with a modified tango, the tango probability is split into two different terms: First, the probability that a standard tango is performed, and second, the probability that the modified tango is performed. The probability of the regular rearrangement remains unchanged. To determine which tango is performed, we need the probability of performing a tango, $P(R5)$ (see equation 4.12), combined with the allocation probabilities of the neighboring storage lanes. The probability of performing the modified tango is the conditional probability of performing a tango, given that there is at least one neighboring storage lane in state E or H. For the analytical derivation, we ignore effects at the edge of the rack, where storage lanes have only one neighboring lane.

To derive the unknown probabilities, consider the following: Each storage lane chosen for retrieval has two neighboring lanes which both can either be in state E, H or F. Hence, nine possible combinations of storage lane states can occur, as presented in Figure 4.15.

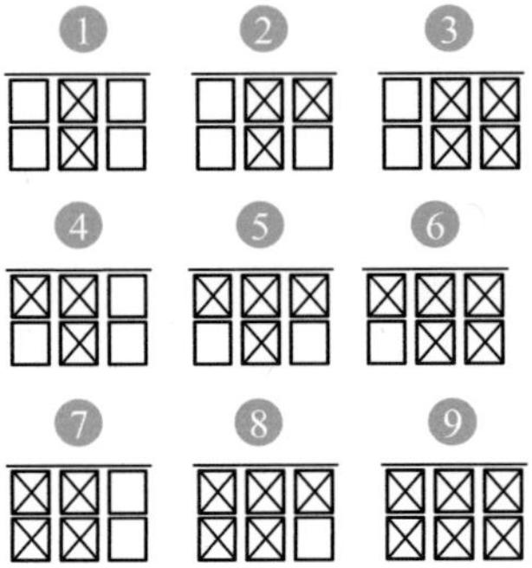

Figure 4.15: Possible combination for the states of the neighboring storage lanes

Eight out of nine states allow to perform a modified tango, while in the remaining a standard tango is performed. The probability of state nine is the probability that a standard tango is performed:

$$P(Standard \mid Tango) = P(F) \cdot P(F) = P(F)^2 \tag{4.68}$$

The probability of performing a modified tango is the probability of the complement, which is:

$$P(Modified \mid Tango) = 1 - P(F)^2 \tag{4.69}$$

Equations 4.68 and 4.69 show how to weight the tango probability.

We have to consider that the modified tango causes state transitions different from those of the standard tango. If a modified tango is performed, the state transitions of the regular rearrangements, i.e., R3 and R4, can be observed. The transition R5 only occurs when a standard tango is performed. Figure 4.16 summarizes this. Allowing for the modified tango, makes R3 and R4 more likely compared to the former model without tango modification and thus affects the probabilities of R3, R4 and R5. A formulation of the changed probabilities is presented in Appendix A. Consequently, the transition probabilities in the Markov Chain change and the stationary allocation of the storage lanes change too. By changing $P(R \mid H)$ and $P(R \mid F)$ in the transition matrix (see equation 4.23), the derivation of the adjusted stationary allocation of the storage lanes is possible. But the solution of the system of equations can only be done numerically and the results are neither appropriately presentable nor applicable. Therefore, we set-up an analytical model for the modified tango by applying the storage lane allocation of the former model using P(E), P(H) and P(F) from equations 4.29, 4.30 and 4.31.

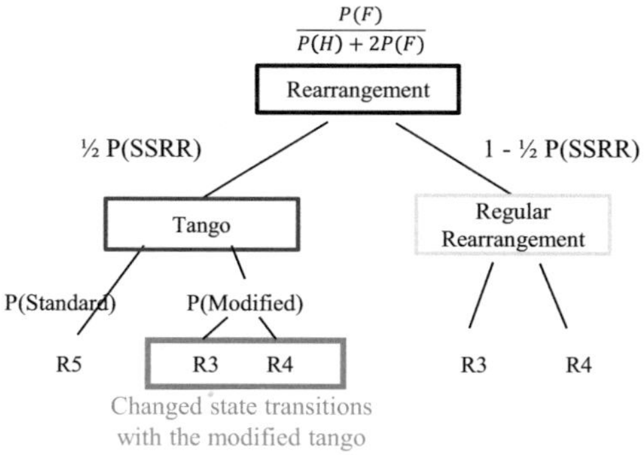

Figure 4.16: Decision tree showing the rearrangement probabilities including both tango variants

The application of the '*old*' probabilities generates an error in the state probabilities of the storage lanes, as the model overestimates the probability of a storage lane being in state H, and underestimates the probability of a storage lane being in state E: For the modified tango, a retrieval of type R3, which is $H \rightarrow E$, can occur in combination with a tango. In the former model, tango always means a transition of type $F \rightarrow H$, which is represented by R4 or R5. When applying the state probabilities of the former model, we therefore automatically ignore R3 transitions that arise due to the modified tango. As a result, the travel time model for the modified tango is slightly inaccurate.

The *approximated* cycle time for the quadruple command cycle with the modified tango is as follows:

$$
\begin{aligned}
E(QC_{dd})_{Mod.Tango} &= t_0 + \frac{5}{2}\cdot\left(\frac{v_x}{a_x}+\frac{v_y}{a_y}\right) + E(QC)^N \cdot \frac{L}{v_x} \\
&+ P(SSRR)\cdot P_{Rearrange}\cdot [P(F)^2 \cdot t_{Tango} + (1-P(F)^2)\cdot t_{Tango,mod.}] \\
&+ (1+(1-P(SSRR))P_{Rearrange}\cdot E(RC)) \\
&+ 4\cdot E(t^S_{LHD}) + 4\cdot E(t^R_{LHD}) - 2\cdot t_{LHD,f}
\end{aligned}
\tag{4.70}
$$

Different from t_{Tango}, the time required to perform the modified tango, $t_{Tango,mod.}$, is stochastic. To determine $t_{Tango,mod.}$, we make the following assumption: If both neighboring lanes are available for a modified tango, the storage lane to perform the rearrangement to is **randomly chosen**. Figure 4.15 illustrates that three cases allow a modified tango to be performed into an empty storage lane (cases 1,3 and 7) and another three cases allow a modified tango to be performed into a half filled storage lane (cases 5,6 and 8). For the remaining cases (2 and 4), both options are possible and thus they are weighted with 1/2 each.

Re-storage	Notation	Probability
Rear position	$P(Mod, rear)$	$\frac{P(E)^2+2P(E)P(F)+1/2\,2\,P(E)P(H))}{(1-P(F)^2)}$
Front position	$P(Mod, front)$	$\frac{P(H)^2+2P(H)P(F)+1/2\,2\,P(E)P(H)}{(1-P(F)^2)}$

Table 4.3: Probabilities of the re-storage position during a modified tango

Based on the probabilities shown in Table 4.3, the time needed to perform a particular modified tango is the standard tango time, t_{Tango}, minus the savings presented in Table 4.2. Consequently, the mean time to perform the modified tango is

$$t_{Tango,mod.} = P(Mod, rear) \cdot (t_{Tango} - t_d) \\ + P(Mod, front) \cdot (t_{Tango} - t_d + (t_{LHD,r} - t_{LHD,f})) \quad (4.71)$$

The idea of the modified tango can be applied to every execution order of the QC. The time of the modified tango, $t_{Tango,mod.}$, is calculated using the respective stationary allocations of the storage lanes (P(E), P(H) and P(F)).

4.5 Validation of the Analytical Model

In this section, we present the validation of the analytical models from the preceding sections by means of a simulation model. First, we introduce our approach, explain how we set-up a comparable configuration and show the results of the validation. Afterwards, we present validation results of a more realistic set-up and analyze rearrangements in particular.

4.5.1 Validation Approach

We apply a simulation model that depicts one storage aisle with an automated S/R machine, having a dual load handling capacity, and double deep storage positions. To implement the model, we use the agent based simulation environment 'AnyLogic'. The simulation allows to model a rectangular rack with any desired number of storage lanes. Within this rack, an AS/RS is modeled that stores and retrieves storage units to and from their positions. The AS/RS moves between the desired positions in the mathematically calculated travel time. Storage and retrieval locations are chosen randomly. Rearrangements that result from the selection of blocked retrieval units are also incorporated in the model. The simulation employs all assumptions of section 4.1.

In order to generate reliable results, we simulate ten replications of different seed values with 100,000 QCs each. Afterwards, the ten results are averaged to obtain the presented numbers. At the beginning of each run, the rack is randomly filled up to the desired filling level. In the individual simulation runs, we do not identify a warm-up period as the influence of this phase is negligible. We conclude this as with a randomly filled rack, the difference of the mean travel time is 0.0006 seconds compared to the case when a warm-up is considered. For a worst case consideration, we can show that starting with unevenly filling the influence on the travel time result is 0.0046 seconds (see Appendix B).

Configuration of the rack and the simulation

The configuration that is applied for the validation is shown in Table 4.7.

Parameter	Setting
Height of the rack	12 m
Length of the rack	24 m
Height of one storage lane	0.04 m
Length of one storage lane	0.08 m
Speed of the AS/RS in x direction	4 m/s
Speed of the AS/RS in y direction	2 m/s
Acceleration of the AS/RS in x direction	∞ m/s^2
Acceleration of the AS/RS in y direction	∞ m/s^2
Shape factor of the rack	1
Time to access the front position	4.5 s
Time to access the rear position	5.5 s
Dead time per LHD cycle	0 s
Mast damping time	0 s
Filling level	90%

Table 4.4: Parameter configuration used to generate results for validation by simulation

Analytical travel time models in general do not fully depict reality, but rather comprise some discrepancies compared to realistic AS/RSs. In contrast, the actual purpose of the simulation is to model realistic scenarios.

Note that these discrepancies are accepted in literature and occur in other established travel time models. In order to improve comparability, we set-up the simulation model to meet the conditions of the analytical model and thus reduce deviations. However, we do not modify the model in general, but rather use a validation set-up that temporarily eliminates these effects. The steps taken in this regard can be distinguished into specific parameter settings, on the one hand, and further modifications of the simulation model, on the other hand. All steps are explained in the following.

Moreover, we set dwell times to zero in both cases, since they are constant in every cycle.

Acceleration and Deceleration: Travel time models assume a full acceleration and deceleration for every movement of the S/R machine. Deviations of travel times between the analytical model and simulation can arise as full acceleration may not be necessary for short travel distances. To eliminate this effect, we assume acceleration or deceleration to be infinite which means that the maximum speed is immediately available. In this way, the travel times for all distances are linearized and consistency of the analytical model and the simulation model is improved.

Continuous storage space: The theoretical, analytical model implicates a continuous storage rack with infinitesimal small storage positions, while every realistic storage rack has discrete storage positions. Consequently, we set the measurements of the storage lanes in a way that many small positions exist. As a trade-off between accuracy and an increased computation time for a larger number of positions, we decide to use tenfold smaller measurements of the storage lanes compared to a realistic size, i.e., the size is modified to $4\,cm$ in height and $8\,cm$ in width.

Height adjustments of the rack: Height and length of the rack give the maximum travel distance assumed for the travel time formula. Since the rack in the simulation model is a non-continuous rack, the AS/RS does not travel to any position but only to discrete, defined storage lanes. To constitute the same maximum travel distance for the AS/RS, we need to adjust the rack in the simulation model: The load handling device is always positioned at the bottom of the particular storage lane it operates at. Hence, the travel distance to the uppermost level is the height of one storage lane less then the total height of the rack. To obtain the same maximum travel distance, we expand the rack of the simulation model by one additional level. The maximum travel distance for the horizontal dimension is adequate, as the I/O position is in the lower left corner of the rack and the S/R machine starts completely outside of the rack. Figure 4.17 illustrates how the load handling device is positioned and the rack is adjusted in the simulation model to correspond to the maximum travel distances from the analytical model.

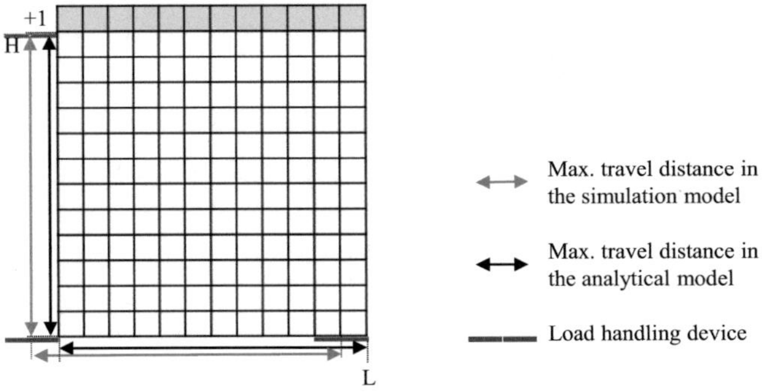

Figure 4.17: Adjustment of the simulated rack to correspond in maximum travel distance

Positioning of the load handling device: The position of the dual load handling device can cause inaccuracies as the choice, which of the load handling devices is used to perform an operation, is not part of the analytical model.

Common travel time models consider standard AS/RSs with one load handling device which is consequently positioned in front of the actual storage position. For two load handling devices, we do not define a fixed order of usage. The decision which device is used next, depends on the execution order of the cycle and the order in which the units are handled. The real travel time can be longer or shorter than the time needed to travel the exact distance between two consecutive locations, if a shift of usage from the left to the right load handling device takes place (or vice versa). Moreover, the direction of movement of the S/R machine influences whether the real travel time is changed. If the vertical travel time is longer than the horizontal travel time, a shift of the LHDs can be performed during the (longer) lifting movement with no change of the travel time.

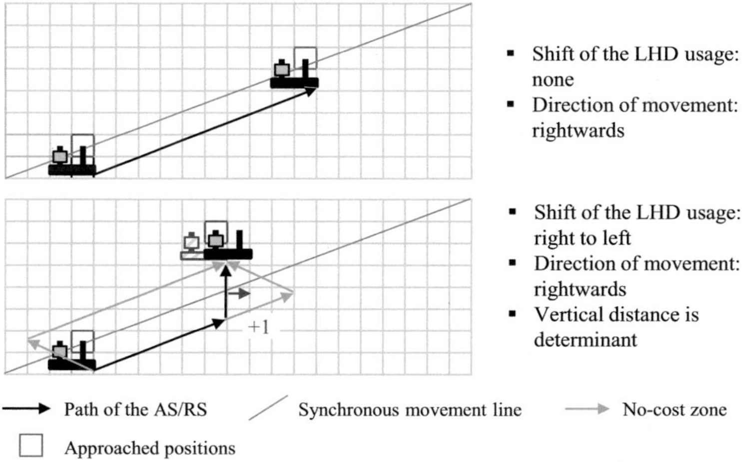

Figure 4.18: On top: A travel movement with no shift. At bottom: The travel time is not influenced by the shift of the LHD because the vertical distance as determinant

Figures 4.18 and 4.19 show these different situations and the resulting change in travel distance compared to a single LHD. Figure 4.18 illustrates two examples where the real travel time corresponds to that of the analytical model. In the upper case, the real travel time is equivalent to that in the analytical model as there is no shift of the LHD usage. The lower case shows that the load handling device can be shifted without influencing the actual travel time due to the longer vertical travel time. In contrast, Figure 4.19 illustrates two cases where a shift of the LHD usage influences the travel time as the horizontal distance is determinant. To be in line with the analytical model, we adjust the simulation to calculate the travel time according to the distance between the approached storage positions only, ignoring the additional time needed for exact positioning of the LHD.

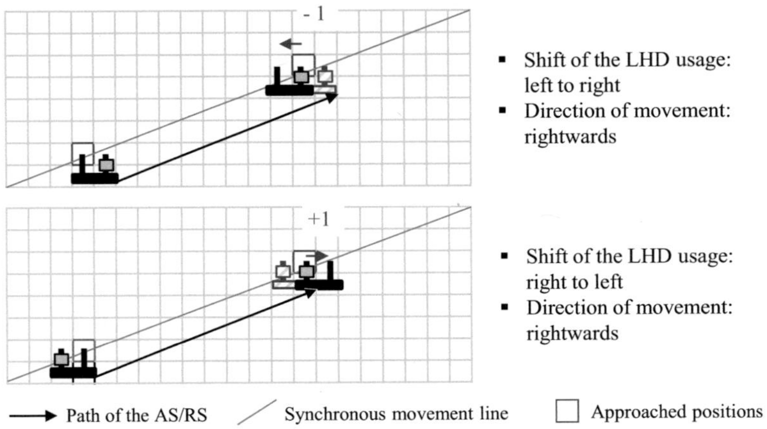

Figure 4.19: Travel distances depending on the shift of the LHD

Before the results are presented, Table 4.5 gives an overview which discrepancies exist and how the simulation is set-up to reduce them.

No.	Discrepancies	Modifications in the simulation
1.	Acceleration and Declaration	Assume to be infinite
2.	Positioning of the dual load handling devices	Calculate travel time without exact positioning in simulation
3.	Discretization of the storage rack	Simulate very small positions

Table 4.5: Different sources of deviations and how they are approached

4.5.2 Validation Result

In Table 4.6 we present the validation results for the basic model according to the explained set-up. We give the 95% confidence interval for the simulation results and absolute as well as the relative deltas between the simulated mean and the computed results.

The travel time delta between the analytical model and simulation is 0.08 seconds, which is a deviation of less than 0.15% from the analytical model. The probabilities for rearrangement and tango show an absolute difference of less than one percent. The probabilities for storage lane allocation exhibit relative deviations of up to more than 2%. Note that the deviations are enhanced due to the high number of considered storage lanes (90300 lanes). When applying storage lanes with a realistic size, the relative delta reaches 1% only.

The deviation in travel time can be explained by two reasons: First, the mean rearrangement distances do not fully correspond in the analytical model and the simulation. We discuss this in more detail at the end of this section. The second reason is that the amount of rearrangements stored in a front position of a storage lane is higher in the simulation than analytically computed. The nearest rearrangement position is found in only a small section of the whole rack, where the overall mean storage lane allocation may not be valid.

4.5.3 Results when comparing to a realistic set-up

As explained, the results from Table 4.6 apply for settings that are not observable in reality. Therefore, we present a comparison where the simulation model represents a realistic AS/RS and the following aspects are set to realistic values:

- Acceleration and deceleration,
- Size of the of the storage lanes,
- The positioning of the load handling device.

The complete list of parameter settings is shown in Table 4.7.

Table 4.8 shows the comparison of the analytical model with the simulation model for the basic model with the new parameter settings. The travel time has a total difference of 0.3059 seconds, which is a delta of about 0.5%.

Reference value	Analytical Model	Simulation Model	95% confidence interval		Delta abs.	Delta rel.
			lower	upper		
Travel time $E(QC)_{dd}$ [s]	54.1173	54.0372	54.0167	54.0577	0.0801	0.15%
$P_{Rearrangement}$	0.3528	0.3527	0.3517	0.3537	0.0001	0.02%
P_{Tango}	0.1176	0.1171	0.1166	0.1176	0.0005	0.43%
% Storage at front	0.6954	0.7031	0.7025	0.7038	−0.0077	−1.11%
% Retrieval from front	0.4704	0.4699	0.4689	0.4709	0.0008	0.17%
% Rearr. at front	0.6954	0.7145	0.7132	0.7158	−0.0191	−2.75%
P(E)	0.0467	0.0456	0.0455	0.0457	0.0011	2.36%
P(H)	0.1066	0.1088	0.1086	0.1089	−0.0022	−2.09%
P(F)	0.8467	0.8456	0.8455	0.8457	0.0011	0.13%
Mean rearr. distance [nl]	1.3396	1.2497	1.2485	1.2509	0.0899	6.71%
Mean rearr. time [s]	9.3582	9.3317	9.3302	9.3332	0.0265	0.28%

Table 4.6: Validation of the analytical model in the validation set-up: Computed vs. simulated results. (The mean rearrangement distance is measured in the number of storage lanes [**nl**])

Parameter	Setting
Height of the rack	12 m
Length of the rack	24 m
Height of one storage lane	0.4 m
Length of one storage lane	0.8 m
Speed of the AS/RS in x direction	4 m/s
Speed of the AS/RS in y direction	2 m/s
Acceleration of the AS/RS in x direction	$3 \, m/s^2$
Acceleration of the AS/RS in y direction	$1.5 \, m/s^2$
Shape factor of the rack	1
Time to access the front position	4.5 s
Time to access the rear position	5.5 s
Dead time per LHD cycle	0 s
Mast damping time	0 s
Filling level	90%

Table 4.7: Realistic parameter configuration used to generate results for validation by simulation

The presented probabilities show an absolute difference of less than one percent, leading to a relative difference of less than 7%.

The mean rearrangement distance exhibits a difference of about 0.05 storage lanes which is accompanied by a difference of 0.91 seconds in the mean rearrangement time for regular rearrangements.

For the SSRR model, the difference in travel time is less than for the basic model. The total difference is 0.1273 seconds which is a delta of 0.202%. The deviations of the other parameters are comparable to those of the basic model or less. The reason for the reduced deviation in travel time is the reduced share of regular rearrangements in the SSRR model, since the observed deviations are mainly caused by rearrangement distance and rearrangement travel time. Detailed validation results for the SSRR model and the model with the modified tango are shown in Tables A.1 and A.2 in the Appendix.

Reference value	Analyt-ical Model	Simu-lation Model	95% confidence interval		Delta abs.	Delta rel.
			lower	upper		
Travel time $E(QC)_{dd}$ [s]	63.3975	63.0916	63.0639	63.1193	0.3059	0.482%
$P_{Rearrangement}$	0.3528	0.3541	0.3534	0.3549	−0.0013	−0.374%
P_{Tango}	0.1176	0.1180	0.1177	0.1183	−0.0004	−0.309%
% Storage at front	0.6954	0.6908	0.6903	0.6913	0.0046	0.662%
% Retrieval from front	0.4704	0.4699	0.4690	0.4707	0.0005	0.110%
Mean rearr. distance [nl]	1.3396	1.2843	1.2835	1.2858	0.0549	4.098%
Mean rearr. time [s]	12.5071	11.5946	11.5932	11.5960	0.9125	7.296%

Table 4.8: Validation of the analytical model: Computed vs. simulated results.
(The mean rearrangement distance is measured in the number of storage lanes [**nl**])

4.5.4 Mean rearrangement distance

As seen in the Tables 4.6 and 4.8, the mean rearrangement distance shows larger differences compared to the other values presented. In Table 4.8 also the mean time needed to perform a rearrangement differs more than other values. The analytical distance is based on the approximation of Lippolt (2003). The resulting mean rearrangement distance depends on the filling level and is outlined in Figure 4.20. The approximation is adjusted such that for a filling level of 0%, the distance is exactly one, because the storage lane next to any random storage lane must be available. As a result of the adjustment, the rearrangement distances decreases slightly between a filling level of 30% and 70%, before it changes to the anticipated behavior of increasing with increasing filling level.

The rearrangement travel time given by formula 4.42 implies a full acceleration and deceleration phase when traveling to the rearrangement position. Up to a filling level of 95%, the mean rearrangement location is less then two storage lanes away (see the following Tables 4.9 and 4.10). In most cases, the S/R machine does not reach its maximum speed in these relatively small distances, which is why the travel time to the rearrangement position is overestimated by equation 4.42.

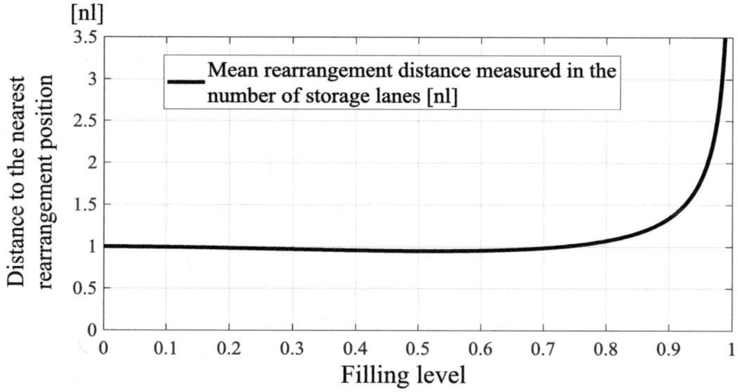

Figure 4.20: Mean rearrangement distance to the next open position according to the approximation of Lippolt (2003) for the values of the validation configuration

In the following, we first compare the **rearrangement distance** from the approximation with exactly calculated distances based on Han et al. (1987) and with simulation results for different filling levels to analyze the quality of approximated and simulated distance (Table 4.9). For the calculation of the **exact distance** we numerically evaluate $E(TB_m)$ (see equation 3.59). Subsequently, we compare the **rearrangement time** from equation 4.42 with calculations of the rearrangement time allowing for the exact values of acceleration and deceleration (Table 4.10). All values presented refer to the validation configuration presented in Table 4.7. Note that for a filling level of 85% the rearrangement distance is below 1. The exact value is calculated in normalized, dimensionless coordinates. To rescale the result, the size of both storage lanes and the rack is required to calculate the distance in number of storage lanes, i.e., exact value [dimensionless coordinates] * length of the rack [m] / length of one storage lane [m/lane].

	Mean rearrangement distance from				
Filling Level	Analytical Model [nl]	Exact Value [nl]	Simulation [nl]	95% confidence interval [nl]	
				lower	upper
0.85	1.165	0.948	1.129	1.128	1.130
0.90	1.340	1.157	1.285	1.284	1.286
0.95	1.777	1.638	1.710	1.709	1.712
0.98	2.702	2.580	2.642	2.639	2.645
0.99	3.652	3.767	3.809	3.804	3.814

Table 4.9: Comparison of rearrangement distances measured in storage lane distances determined by analytical model, exact method and simulation.

Except for a filling level of 99%, the results of the simulation lie between those of the analytical model and the exact value (see Table 4.9) for all filling levels. Thus, the simulation results have the right dimension. The positive difference of analytical minus simulation value explains both positive deviations observed in total travel time as well as overestimation of the rearrangement travel time.

Filling Level	Travel time according to formula 4.42 [s]	**Exact** travel time for the approximated distance [s]	**Exact** travel time for the **exact distance** [s]	**Exact** travel time for the simulated distance [s]
0.85	1.566	1.115	1.006	1.098
0.90	1.601	1.195	1.111	1.171
0.95	1.689	1.377	1.322	1.351
0.98	1.874	1.689	1.659	1.679
0.99	2.088	2.006	2.005	2.016

Table 4.10: Comparison of the one-way travel time measured in seconds to the rearrangement position for analytical model, exact method and simulation.

Based on the different distances from Table 4.9, Table 4.10 contrasts the one-way rearrangement travel time according to equation 4.42 with the exact travel times. **Exact** describes the application of the triangular velocity profile (see Figure 3.1) with real amounts of acceleration and deceleration. Table 4.10 shows that formula 4.42 provides significant deviations, especially when compared to the deviations between the set of all exact calculated values. We observe that both, the exact travel time based on the approximated distance (column 3) as well as the exact travel time derived by the simulation (column 5) only slightly differ from the exact travel time (column 4), as mostly all deviations are less than 0.1 seconds. We conclude that both distances are sufficiently accurate. Apparently, the deviations in the validation results mainly arise from the overestimation caused by equation 4.42. Table 4.10 shows that the deviations decrease with increasing filling level. The travel time according to formula 4.42 only shows a deviation of 0.082 seconds at 99% filling level. This is explained by the growing relevance of acceleration and deceleration for longer rearrangement distances.

For storage lanes at the edge of the rack, the mean-value based rearrangement approach does not hold. For these positions, the nearest rearrangement location is further away than for any position inside the rack. The extent of this effect is depending on the ratio of the number of storage positions at the edge of the rack and the total number of storage positions. In contrast to the computed distances, the simulated distances contain this effect. However, since all **exact** travel times differ less than 0.1 seconds, we conclude that deviations mainly arises due to acceleration and deceleration rather than from the travel distances themselves (Table 4.10). Moreover, this is in line with the distinct deltas observed for the mean rearrangement time when comparing Table 4.6 and Table 4.8.

4.6 Comparisons and Discussions

In this section we compare and evaluate our approach. First, we compare the models employed in this thesis and study further details of their characteristics. In a second step, we compare our models to traditional travel time models discussed in literature. Unless specified otherwise, the calculations are based on the configuration mentioned in Table 4.7. Dwell times are set to $t_{dead} = 0.3s$ and $t_{mast} = 1s$ different from Table 4.7.

4.6.1 Comparison of the derived models

We compare the presented models in terms of travel time, allocation of the storage lanes and rearrangement characteristics.

Figure 4.21 shows the surface representing the travel time of the general model depending on P(SSRR) and the filling level. The surface representing the travel time for the model with the modified tango (formula 4.70) shows the same behavior and is not shown. The total difference in travel time is small for varying P(SSRR). In general, we observe the longest travel times for P(SSRR) = 0 and the shortest travel times for P(SSRR) = 1. The difference arises from changed mean load handling times for storage and retrieval as well as changed rearrangement and tango effort. While the difference from load handling times is minor, the differences in rearrangement times, both regular rearrangements and tango, makes up the largest part of the total difference observed. The reason for this behavior is the greater amount of tango rearrangements, occuring with increasing P(SSRR).

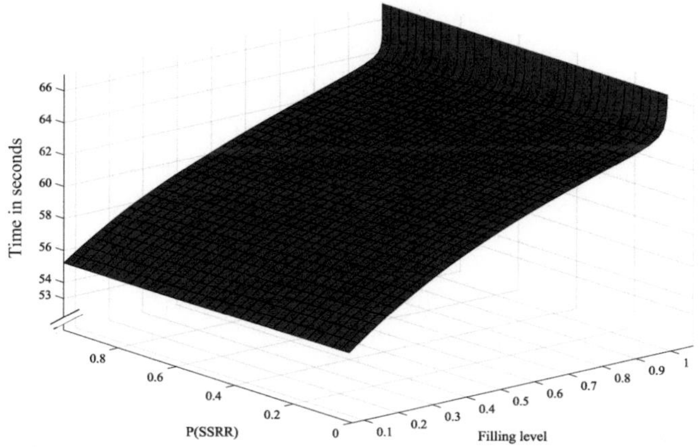

Figure 4.21: Travel time in seconds depending on P(SSRR) and the filling level

An increasing filling level causes an increase of the rearrangement time, while the tango time remains unchanged. This explains the constantly increasing difference in travel time between the limit values of P(SSRR)= 0 and P(SSRR)= 1 for a raising filling level. Consequently, the difference is greatest for highest filling levels.

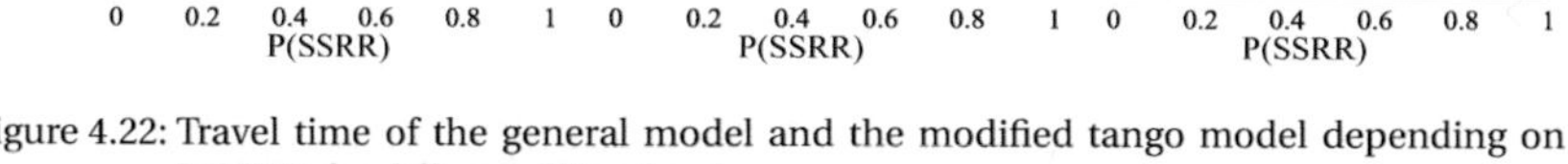

Figure 4.22: Travel time of the general model and the modified tango model depending on P(SSRR) for different filling levels

Figure 4.22 shows a comparison of the travel time for the general model and the modified tango model over P(SSRR) and for three different cases of filling levels. The travel time of the model with the modified tango consistently lies below the travel time of the general model. With an increasing amount of tango (increasing P(SSRR)), the travel time difference to the general model increases. However, the maximum difference gets smaller the higher the filling level becomes. While for small and medium filling levels, the chance to perform the modified tango is relatively high, it lessens with increasing filling level, causing the convergence of the modified model and the general model.

On a quantitative basis, the general model performs best with increasing amount of tango rearrangements, i.e., increasing P(SSRR), for the chosen parameter configuration. However, differences in travel time are rather small and will not create significant variances in throughput, which is why we subsequently discuss whether a greater amount of tango is always preferable and should be aimed for.

Evaluation for changed parameters: Role of the Tango

For P(SSRR) = 1, shorter travel times are observed due to smaller rearrangement times. Figure 4.23 illustrates the mean rearrangement time, both for the general model with P(SSRR) = 1 and P(SSRR) = 0, as well as the tango time. For all filling levels, the time needed to perform a tango rearrangement is smaller than any rearrangement time.

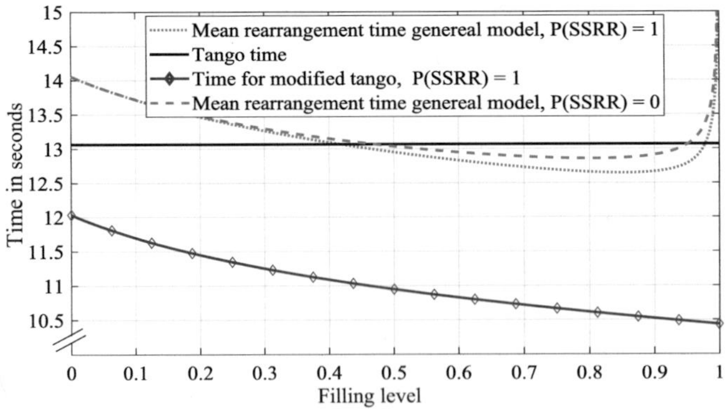

Figure 4.23: Comparison of mean rearrangement time and tango time for basic and SSRR model

With such constellations, an operating strategy with a greater amount of tango leads to shorter travel times. With the modified tango, this situation is further enhanced. However, whether a higher amount of tango rearrangements leads to an improved performance is dependent on specific configuration parameters. Important drivers in this context are

- The access time to the rear position ($t_{LHD,r}$).
- The rearrangement distance, which itself is depending on the filling level and the stationary allocation of the storage lanes.

To explain the first driver, consider a (rather slow) load handling device, which needs twice the time to access the rear storage position than the front storage position, hence $\frac{t_{LHD,r}}{t_{LHD,f}} = 2$. In this case, tango is disadvantageous, because the re-storing is always performed to the rear position of the storage lane, while regular rearrangements allow re-storing into a free front position as well. Additionally, for high filling levels, the ratio of rearrangements into front positions increases. This situation can be analyzed depending on the state of a storage lane that is chosen for rearrangement, on the one hand, and the time needed to travel to the respective position on the other hand.

State of the storage lane chosen for rearrangement	Condition for favor of tango
E	$(E(RC)/2) - t_d > t_{LHD,r} - t_{LHD,r}$
H	$(E(RC)/2) - t_d > t_{LHD,r} - t_{LHD,f}$

Table 4.11: General condition for tango to have an advantage over regular rearrangements for the different states of the rearrangement storage lane

The first inequation in Table 4.11 is always true. Compared to rearrangements into an empty storage lane, a tango is always beneficial, as the time needed to travel to the rearrangement position ($E(RC)/2$) can not fall below the positioning time the LHD requires during the tango (t_d). For a half-filled storage lane, it depends on the **difference** between the **access times** to the **front** and the **rear** position of a storage lane as well as the time needed to travel to the chosen rearrangement position. The larger the difference

$_{LHD,r} - t_{LHD,r}$, the higher is the chance a regular rearrangement is preferable over a tango. Figure 4.24 and 4.25 illustrate the comparison in further detail. Remember, for high filling levels, the ratio of rearrangements into half-filled storage lanes increases. Figure 4.24 shows the comparison applied in Figure 4.23 with a changed access time to the rear position. $t_{LHD,r}$ is increased by one second to 6.5 seconds, whereas all other parameters remain unchanged. We see, in this situation the tango rearrangement is only beneficial for extreme filling levels (>95%). In Figure 4.25, this comparison is generalized by showing the same terms depending on the ratio $\frac{t_{LHD,r}}{t_{LHD,f}}$ as a variable with the filling level set to 90%. The ratio in the changed situation ($t_{LHD,r} = 6.5$) is $1.44 (= 6.5/4.5)$, while in the former configuration it was 1.22. This shows that the benefit of the tango depends on the load handling times.

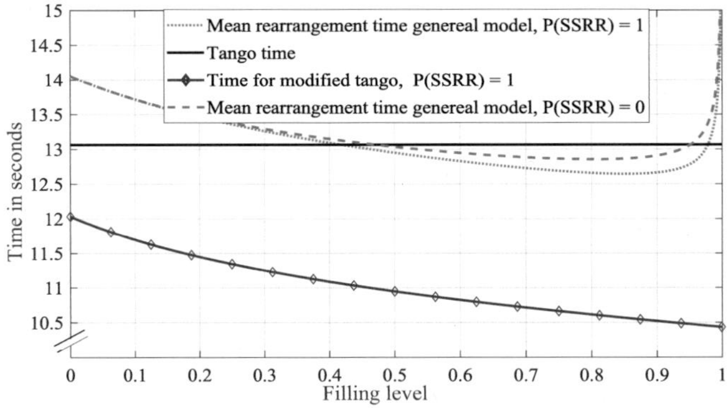

Figure 4.24: Mean times for rearrangement and tango with increased $t_{LHD,r}$ of 6.5 seconds

Utilization of the modified tango makes the tango's benefit more resilient towards an increased ratio of $\frac{t_{LHD,r}}{t_{LHD,f}}$. Figure 4.25 shows that the mean time of the tango rearrangement in the modified tango model is more robust with increasing relations of $\frac{t_{LHD,r}}{t_{LHD,f}}$ than the standard tango. The purple line intersects the lines for regular rearrangements later, i.e., for higher ratios, than the black line. Recall the example of the slow LHD with a ratio of $\frac{t_{LHD,r}}{t_{LHD,f}} = 2$. In this situation, both tango variants are outperformed by regular rearrangements, as we can observe both tango lines being above the rearrangement lines in Figure 4.25.

However the quantitative result may not be the only fact considered when addressing the employment of tango. Tango offers the following additional advantages that are independent of the discussed parameters:

- Tango is **independent of storage assignment policies**. For tango rearrangements, no additional rearrangement position is needed. This is of increasing interest for very high filling levels or in cases in which an available position for a rearrangement may be further away than the mean analytical distance, e.g. for turnover based storage assignment, with SKU identical lanes or any kind of dedicated storage.
- Tango causes **horizontal movements** only. Regular rearrangements require horizontal and vertical movements and thus are more energy-consuming, as lifting in general needs more energy than driving (Braun 2016, p.302 f.).
- Tango produces more **half-filled storage lanes** on average. This is beneficial for the overall performance, as half-filled storage lanes serve as potential rearrangement positions and prevent rearranging. This fact is already included in the quantitative results and therefore also in our analysis. However it may effect other policies that require potential storage lanes, e.g. some of the operating policies discussed in the following chapter.

QC with greater amount of **tango** appear **most efficient**, especially when including the modified tango and the additional advantages. We conclude that it is generally favorable to allow for the greatest possible amount of tango.

Transfer of the models to other LHD configurations

To end this section we shortly discuss the relaxation of the LHD's design condition. In section 4.1.1 we formulate the assumption that the LHDs are horizontally arranged, having the same distance between each other as two storage lanes.

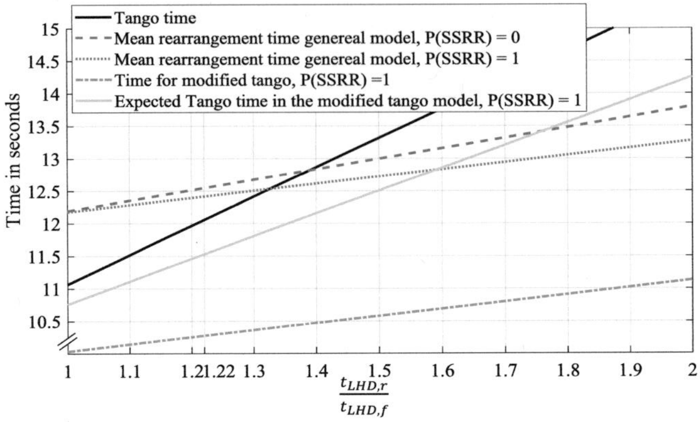

Figure 4.25: Mean times for rearrangement and tango depending on the relation of $t_{LHD,r}$ and $t_{LHD,f}$ for a filling level of 90 %

Vertical Arrangement of the LHDs A possible relaxation of the LHD design is a vertical arrangement of the two load handling devices above each other. All models presented in this chapter are still valid with this assembly. Tango is still possible, where the difference is that the LHD moves vertically instead of horizontally. Figure 4.26 shows the lateral view of the tango for this case. If a storage lane either above or below the lane of retrieval unit has an empty position, the modified Tango is possible, which is why also that model is valid.

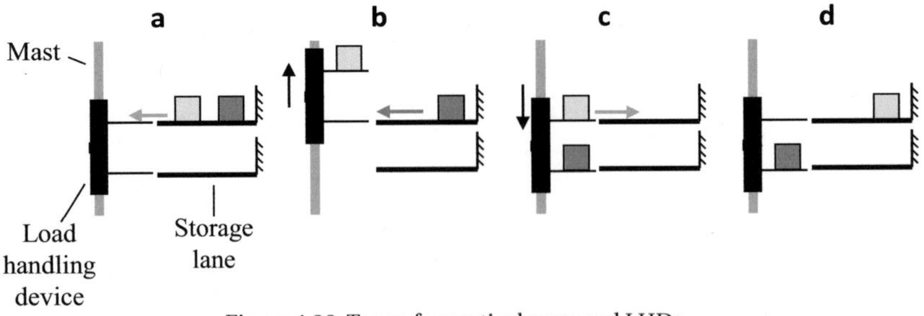

Figure 4.26: Tango for vertical arranged LHDs

A potential drawback of this setting is that for every tango lifting movements are required. On the contrary, vertically arranged LHDs exert less torsion to the mast.

Varying distance between the LHDs A possible relaxation is that the distance between the two load handling devices is changed. The QC in the general model is not affected by this change and Tango is possible. The difference is that the time needed to perform a Tango changes according to the actual distance between the LHDs. Figure 4.27 shows how a tango rearrangement works in this case.

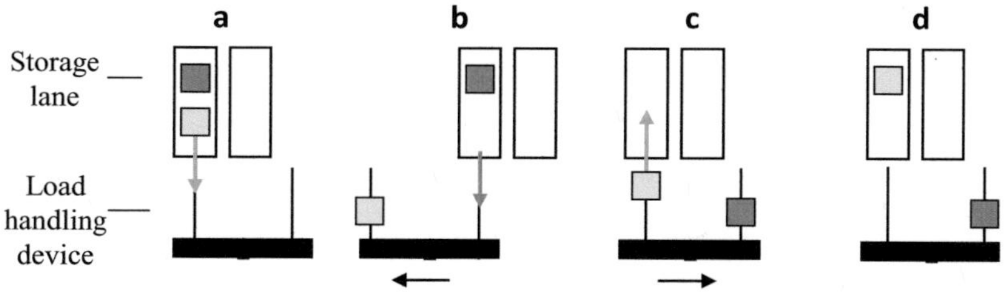

Figure 4.27: Tango for changed distance between LHDs

We can see that only one of the LHDs is positioned in front of a storage lane. As a consequence, the modified Tango is not possible. The same implications hold for the case of vertically arranged LHDs with a changed distance between each other.

4.6.2 Comparison to previous work

We compare the presented models to already existing work by two different approaches.

1. Comparing our model to **traditional** travel time models such as travel time formulas for dual command cycles in double deep storage areas. According to Meller and Mungwattana (1997), we divide the time of the quadruple command cycle by two in order to allow a comparison

to dual command cycles. Note that in dual command cycles only two units are handled.

2. Comparing our model to more **similar approaches** that also try to determine a quadruple command cycle in a double deep storage environment. We use Xu et al. (2015) and Kayser (2003) for this purpose.

Comparison to dual command cycles

The travel time for a random dual command cycle in a double deep storage rack (Lippolt 2003) is indicated with $E(DC)_{dd}$.

Value of comparison	Total travel time
$E(QC)_{dd}/2$	31.6987s
$E(DC)_{dd}$	39.6367s

Table 4.12: Travel time comparison of the basic model and the dual command cycle for a filling level of 90%

The comparison which is illustrated in Table 4.12 as well as in Figure 4.28 shows an increased throughput compared to the dual command cycle.

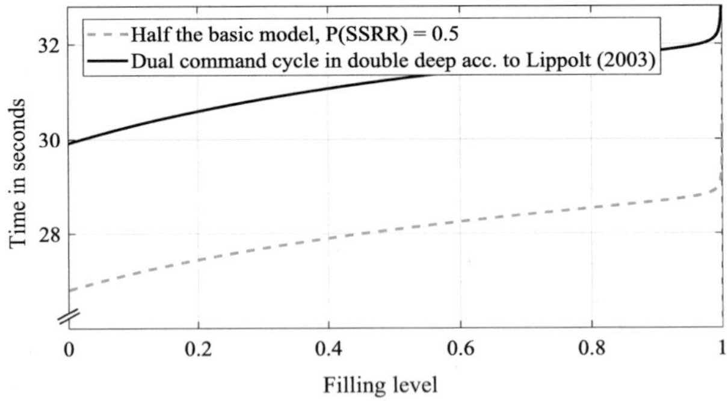

Figure 4.28: Travel time comparison of the basic model and the dual command cycle according to Lippolt (2003)

Meller and Mungwattana (1997) report an improvement of the quadruple command cycle over the dual command cycle of 24% for single-deep storage systems, while they neither consider acceleration and deceleration nor load handling times. In the double deep case, we find an improvement of 20% compared to the appropriate dual command cycle. Without taking acceleration and deceleration as well as load handling times into account, the value is increased to 22%. 24% are not reached due to rearrangement efforts. In rearrangements, no time is saved in a quadruple command cycle as rearrangements in a double deep setting affect the travel time of the dual command cycle in the same way they do it for the quadruple command cycle.

Comparison with quadruple command cycles

The model of Kayser (2003) ($E^{dt,nl}(tvs)$) which is very similar to the approach of Lippolt (2003) assumes a fixed SSRR sequence but does not incorporate a tango rearrangement. State probabilities of the storage lanes are adopted from Lippolt (2003). From Xu et al. (2015) we choose the first model for comparison ($E(QC)^{FCFS}$). Although their second, modified model ($E(MQC)$) is more consistent to our assumptions as storage lanes can attain all three states, their modified model requires the number of half filled storage lanes as an input factor, which imposes restrictions for application in practice. Moreover, they assume a successive multiple storage of both units into an empty storage lane at the first storage operation, an assumption that does not correspond to the random execution of our model. As their model is only defined for filling levels greater than 0.5, we only calculate values of $E(QC)^{FCFS}$ in this range. Furthermore, the following adjustments are made to the model of Xu et al. (2015) to ensure a reasonable comparison:

1. Acceleration and deceleration are incorporated.
2. Twice the access time to the front position ($t_{LHD,f}$) is added.

The comparison is illustrated in Table 4.13 as well as in Figure 4.29.

The model of Kayser (2003) does not differ significantly from our model: Consistently, the obtained travel times are only slightly above those of the basic model with P(SSRR) = 0.5. As the travel time for P(SSRR) = 1 is less, we can not derive an advantage of using the dual load handling capacity to shorten the rearrangement cycle as assumed by Kayser (2003). The travel time calculated based on the model of Xu et al. (2015) is below those of the basic model and the model of (Kayser 2003) until filling levels of around 60%. Afterwards, the travel time increases significantly with a raising filling level. The observed behavior suggests our model to be more robust for higher, practice-oriented filling levels.

Value of comparison	Total travel time
$E(QC)_{dd}$	63.3975s
$E^{dt,nl}(tvs)$ (Kayser 2003)	63.4688s
$E(QC)^{FCFS}$ (Xu et al. 2015)	65.7911s

Table 4.13: Travel time comparison to other models for quadruple command cycles for a filling level of 90%

Concluding the discussions of this chapter, we have derived a general model for travel time determination of a QC that incorporates all possible execution orders. Besides, we have provided an extensive evaluation of the model by successfully validating it via a simulation and comparing it to existing models from literature.

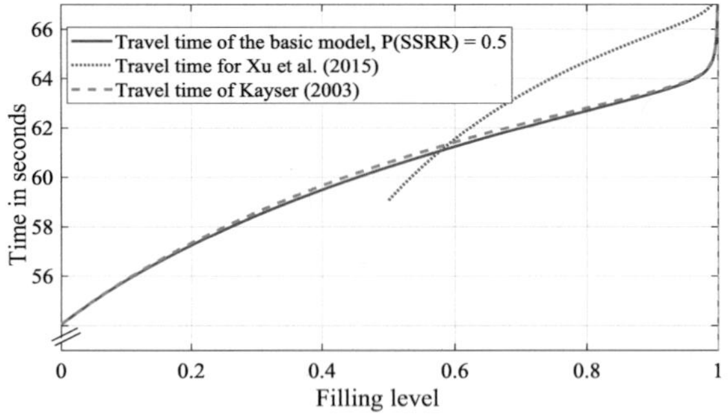

Figure 4.29: Comparison of performance to other models for quadruple command cycles according to Xu et al. (2015) and Kayser (2003)

5 Strategies for Improved Throughput

The best way to find yourself
is to lose yourself in the
service of others.
-M. Gandhi

In the preceding chapter we have presented a general model assuming a random execution of the quadruple command cycle. This is necessary to provide a quantitative basis in accordance with established models for single command and dual command cycles. Moreover, the general model can serve as a benchmark for further types of execution with the objective of improving cycle times. The dual capacity of the load handling device offers a greater level of freedom in operation compared to traditional AS/RSs. In this chapter, we present different strategies to perform the quadruple command cycle with the objective of improving the cycle time while the setting of the system remains unchanged.

5.1 Methodology of Strategy-Definition

Figure 5.1 illustrates the approach for the definition of strategies. As it is based on a literature review, we examine publications concerning AS/RSs with dual- or multiple- load handling devices that formulate routing and sequencing heuristics for the execution of the related command cycles. While not inevitably dealing with double deep storage, most ideas for single deep storage presented in these papers can also be applied in a double deep storage environment. We formulate strategies by composing elements of previously used approaches that are relevant for our investigations. The most important publications are summarized in Table 5.1. In a last step, we structure and visualize all strategies in tabular form. Table 5.2 is the result of this process and represents the long list of all strategies defined. Besides, the results of the two last steps are similar to a morphological box.

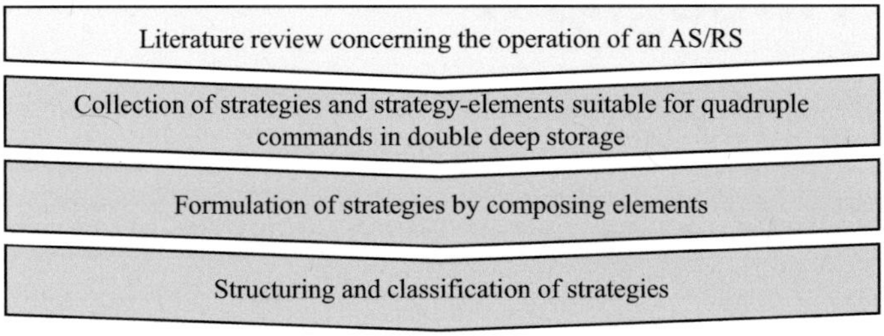

Figure 5.1: Steps in strategy definition

To start our explanations, we define what we refer to as a **Strategy** and which elements a strategy consists of in our approach. Crucial for the definition of a command cycle is the logic applied for selecting all storage and retrieval requests that are processed. The elements that are used to describe a strategy are summarized at the top of Figure 5.2. At the same time, they represent the different parameters of the morphological box. In contrast to a conventional morphological box, the different parameters and their attributes are not independent of each other. Nevertheless, a strategy can be described as the realization of attributes (one or multiple) for every parameter, i.e. the elements of a strategy.

Storage policies, retrieval policies as well as routing and sequencing rules closely interact with each other and can not be addressed independently (Kraul (2010), Roodbergen and Vis (2009)).

We distinguish five elements (parameters) of a strategy belonging to the concept of **routing and sequencing** as explained in section 2.2.

1. The **Main routing policy** is the first priority for the execution and is derived from the most common concepts of operating an AS/RS as

shown in Table 5.1. The main routing policies we distinguish are the following: Shortest Leg, Nearest Neighbor, Flip Flop, Multiple storage, Increase Tango and Class based storage. Moreover, these policies represent the main category in the classification of Table 5.2.

Author	Nearest Neighbor	Shortest Leg	Flip Flop	Multiple	Class based
Eynan and Rosenblatt (1993)	x				x
Grafe (1997)		x		x	x
Graves et al. (1977)					x
Han et al. (1987)	x	x			
Keserla and Peters (1994)	x		x		
Kraul (2010)		x			
Meller and Mungwattana (1997)	x	x			
Sarker et al. (1991)	x		x		
Sarker et al. (1994)			x		x
Seemüller (2006)	x			x	
Xu et al. (2015)	x			x	

Table 5.1: Referencing of the used strategy elements in literature

2. **Additional rules** basically comprise the same concepts like the main routing policies. They are used to complement the main routing policy for those storage and retrieval locations that are not yet defined by the main routing policy.
3. The **Execution order** determines whether the order of the quadruple command cycle is SSRR or SRSR.
4. There are two possibilities for **Storage selection** which are a fixed, i.e. predefined, or a free selection of the storage location. With free selection, having more than one possible open location, the exact location is selected with respect to the storage assignment policy (see 6.).
5. **Retrieval selection** covers two aspects (retrieval policy and resorting of the retrieval list) as defined in subsection 2.2.2. For the remain-

der of this chapter, we reduce these two aspects to either fixed or free selection of retrieval requests. Strategies determine whether free retrieval selection as a characteristic of the AS/RS is required.

The main routing policies and the additional rules are explained in subsection 5.2.2 in more detail. Next to these five major elements, two additional elements are used to describe strategies:

6. **Storage assignment** is the decision which storage positions are permitted for a given type of SKU (see subsection 2.2.1).
7. **Rearrangement rules** describe all policies related to rearrangements.

Both (6. and 7.) are not part of Table 5.2, because storage assignment is a separate decision and rearrangement rules do not vary. We explain their influence in subsection 5.2.2 in more details.

Dwell point policies, idle time policies or batching of requests are not considered in our approach. The morphological box (Figure 5.2) also illustrates the process of defining the strategies in three steps: In a **first step**, the main routing policy is chosen. It defines the **main category** the strategy belongs to in Table 5.2 and is presented in column one. Afterwards, further options for storage and retrieval are added from column two. The options chosen in column two must not counteract the logic of the main routing policy and only involve storage and retrieval requests not concerned by the first step. Finally, the remaining elements of routing and sequencing are evaluated for the given selection of attributes.

Note that due to the dependencies within routing and sequencing, the chronological order of selecting the characteristics is important as described above. The main routing policies of the strategy is the first priority for the execution and serves as the main category. Main routing policies determine to which extent the execution of the cycle is already defined and how many additional rules can be added in **step two**. Consider the following example: For all strategies of the main routing policy *Flip Flop*, by definition the second storage operation is performed into the position of the first retrieval. As a consequence, the first storage and the second retrieval remain 'undefined' and can be specified by additional rules.

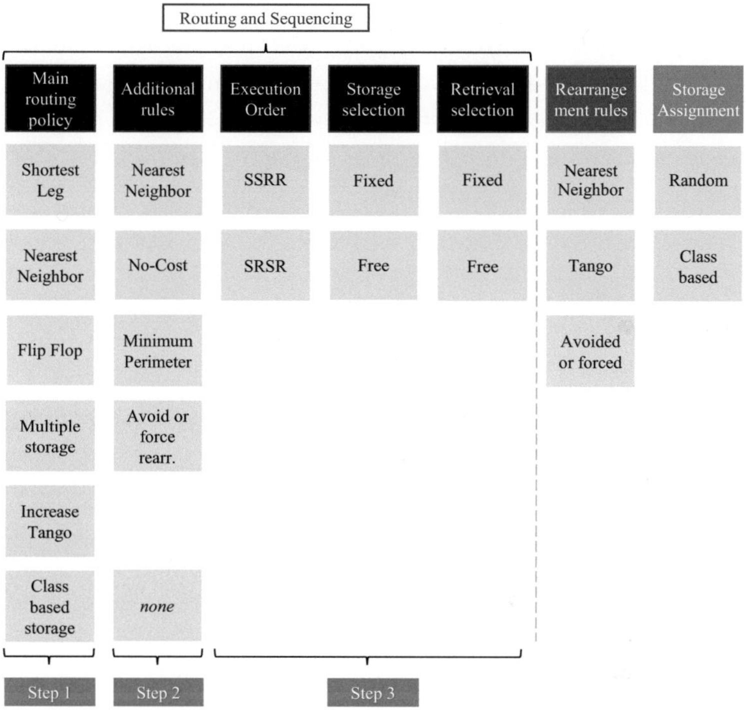

Figure 5.2: Morphological box describing the elements and their attributes for defining strategies

Likewise, **step three** is dependent on the rules defined in step one and two. We differentiate between fixed and non-fixed positions for the selection of both storage and retrieval positions. **Fixed** positions mean, the locations are randomly chosen prior to each cycle and can not be chosen freely. This is also referred to as *predefined*. On the contrary, in the non-fixed case, positions can be **selected freely** in an appropriate way for the given strategy. The first and second step define which way of storage and retrieval selection is required. Recall the Flip Flop example to clarify how the third step is influenced:

Due to the Flip Flop movement, a free selection for storage positions is required in order to allow storing into the retrieval position. The main routing policy, Flip Flop, itself defines no rule for the first storage position, which is why it is freely, but randomly chosen. As no specifications are made for the retrieval positions, free selection is not required by Flip Flop and thus retrieval positions can be predefined.

However, if a specific strategy has no requirement for free selection of retrieval positions, generally both options (fixed or free) of retrieval selection are possible. On the contrary, if free selection of retrieval positions is required, a fixed selection of retrieval positions does not work. Therefore, the attributes that can be chosen in step two and three are depending on preceding steps. An overview of all compatibilities can be found in the Appendix A (Figure A.1).

The complete list of strategies, presented in Table 5.2, is developed by assigning the different attributes from all elements to every main routing policy when possible.

5.2 Presentation and Explanation of all Identified Strategies

In section 5.2.1 all strategies are presented in tabular form (Table 5.2). While we do not describe every single strategy in detail, the remaining characteristics are explained in subsection 5.2.2.

5.2.1 The Longlist

ID	Main routing policy	Additional rules and description	Selection of the single requests				Execution order	Choose storage position freely	Choose retrieval requests freely
			SP1	SP2	RP1	RP2			
SL1	*Shortest Leg*	no additional rules	random	on the way to R1	fixed	fixed	SSRR	x	-
SL2	*Shortest Leg*	+ RNN for retrievals	random	on the way to R1	fixed	fixed	SSRR	x	x
SL3	*Shortest Leg*	+ retrieval near retrieval	random	on the way to R1	random	next to R1	SSRR	x	x
SL4	*Shortest Leg*	+ using No-Cost Zone	random	on the way to R1	random	in No-Cost between R1 and I/O	SSRR	x	x
SL5	*Shortest Leg*	+ avoid regular rearrangements	random	on the way to R1	blocked one	non-blocked one	SSRR	x	x
NN1	*Nearest Neighbor*	simple	fixed	fixed	fixed	fixed	SSRR or SRSR	-	-
NN2	*Nearest Neighbor*	RNN	fixed	fixed	fixed	fixed	SSRR or SRSR	-	-

ID	Main routing policy	Additional rules and description	Selection of the single requests				Execution order	Choose storage position freely	Choose retrieval requests freely
			SP1	SP2	RP1	RP2			
NN3	*Nearest Neighbor*	RNN + simple	fixed	fixed	fixed	fixed	SSRR or SRSR	-	-
NN4	*Nearest Neighbor*	closest storage in x-direction, second storage, closest retrieval in x-direction, second retrieval	fixed	fixed	fixed	fixed	SSRR	-	-
NN5	*Nearest Neighbor*	storage near retrieval	next to R1	next to R2	fixed	fixed	SRSR	x	-
NN6	*Nearest Neighbor*	storage near retrieval + RNN for retrievals	next to R1	next to R2	fixed	fixed	SRSR	x	-
NN7	*Nearest Neighbor*	retrieval near storage	fixed	fixed	next to S1	next to S2	SRSR	-	x
NN8	*Nearest Neighbor*	retrieval near storage + RNN for storage	fixed	fixed	next to S1	next to S2	SRSR	-	x
NN9	*Nearest Neighbor*	pairing	pair up next to a R	pair up next to a R	pair up next to an open position	pair up next to an open position	SRSR	x	x

ID	Main routing policy	Additional rules and description	Selection of the single requests				Execution order	Choose storage position freely	Choose retrieval requests freely
			SP1	SP2	RP1	RP2			
FF1	*Flip Flop*	no additional rules	random	into position of R1	fixed	fixed	SRSR	x	-
FF2	*Flip Flop*	+ simple NN for retrievals	random	into position of R1	fixed	fixed	SRSR	x	-
FF3	*Flip Flop*	+ RNN for retrievals	random	into position of R1	fixed	fixed	SRSR	x	-
FF4	*Flip Flop*	+ storage near retrieval	next to R1	into position of R1	fixed	fixed	SRSR	x	-
FF5	*Flip Flop*	+ simple NN for R2	random	into position of R1	random	next to R1	SRSR	x	x
FF6	*Flip Flop*	+ storage near retrieval + simple NN for R2[1]	next to R1	into position of R1	random	next to R1	SRSR	x	x
FF7	*Flip Flop*	+ using No-Cost Zone	random	into position of R1	random	in No-Cost between R1 and I/O	SRSR	x	x

[1] see Sarker, Sabapathy, Lal and Han (1991)

ID	Main routing policy	Additional rules and description	Selection of the single requests				Execution order	Choose storage position freely	Choose retrieval requests freely
			SP1	SP2	RP1	RP2			
FF8	*Flip Flop*	+ storage near retrieval + using No-Cost Zone	next to R1	into position of R1	random	in No-Cost between R1 and I/O	SRSR	x	x
FF9	*Flip Flop*	+ Minimum Perimeter	next to R1	into position of R1	random	next to R1	min (SR1R2; SR2R1)	x	x
FF10	*Flip Flop*	+ Minimum Perimeter + No-Cost Zone	next to R1	into position of R1	random	in No-Cost between R1 and I/O	min (SR1R2; SR2R1)	x	x
MS1	*Multiple-Storage*	no additional rules	random[2]	random	fixed	fixed	SSRR	x	-
MS2	*Multiple Storage*	+ simple NN for retrievals	random	random	fixed	fixed	SSRR	x	-
MS3	*Multiple Storage*	+ RNN for retrievals	random	random	fixed	fixed	SSRR	x	-
MS4	*Multiple Storage*	+ storage near retrieval	next to R1	next to R1	fixed	fixed	SSRR	x	-

[2] For all *Multiple Storage* strategies, both variants are possible: Either simultaneous storing in two lanes next to each other or successive storing of both jobs in one empty storage lane. This means storage positions are chosen randomly among all neighboring pair of available positions or among all empty storage lanes.

ID	Main routing policy	Additional rules and description	Selection of the single requests				Execution order	Choose storage position freely	Choose retrieval requests freely
			SP1	SP2	RP1	RP2			
MS5	*Multiple Storage*	+ using No-Cost Zone	random	random	random	in No-Cost between R1 and I/O	SSRR	x	x
MS6	*Multiple Storage*	+ using No-Cost Zone	random	random	in No-Cost between S and R2	random	SSRR	x	x
MS7	*Multiple Storage*	+ retrieval near storage	random	random	next to S	next to R1	SSRR	x	x
MS8	*Multiple Storage*	+ retrieval near storage + using No-Cost Zone	random	random	next to S	in No-Cost between R1 and I/O	SSRR	x	x
MS9	*Multiple Storage*	+ avoid regular rearrangements	random	random	blocked one	non-blocked one	SSRR	x	x
IT1	*Increase Tango*	+ avoid regular rearrangements	fixed	fixed	blocked one	non-blocked one	SSRR	-	x
IT2	*Increase Tango*	+ avoid regular rearrangements + RNN for storage	fixed	fixed	blocked one	non-blocked one	SSRR	-	x
IT3	*Increase Tango*	+ avoid regular rearrangements + retrieval near storage	fixed	fixed	blocked one	non-blocked one next to R1	SSRR	-	x

ID	Main routing policy	Additional rules and description	Selection of the single requests				Execution order	Choose storage position freely	Choose retrieval requests freely
			SP1	SP2	RP1	RP2			
IT4	*Increase Tango*	+ avoid regular rearrangements + using No-Cost Zone	fixed	fixed	blocked one	non-blocked one in No-Cost between R1 and I/O	SSRR	-	x
SL+ MS+	*Multiple Storage Shortest Leg*	for strategies with SSRR sequence and remaining degrees of freedom for retrievals + former *Increase Tango* strategies	random	random	blocked one	non-blocked one +ITx	SSRR	x	x
CB1	*Class based storage*	retrieval from same class as storage	random	random	from class of S1 or S2	from class of S1 or S2	SSRR or SRSR	x	x
CB2	*Class based storage*	retrieval near storage; same class	random	random	next to S1 from same class	next to S2 from same class	SRSR	x	x

Table 5.2: Collection of identified strategies for quadruple command cycle in double deep storage environments

5.2.2 Explanation of the Strategy Components

In this subsection we describe how the main routing policies and the additional components defining the strategies, listed in Table 5.2, work.

Main routing policies

Shortest Leg The objective of the *Shortest Leg* heuristic is the selection of a storage position along the way to a retrieval position from the no-cost zone, causing no additional travel time (Han et al. 1987). In our adaption for the quadruple command cycle, the second storage is performed according to this procedure, while the execution order SSRR is applied. This means, the first storage position as well as the first retrieval position are randomly selected and define the travel path of the S/R machine. In contrast to Han et al. (1987), this prevents the development of a higher filling level close to the I/O point, as the first storage position is selected randomly (see Kraul (2010)).

Nearest Neighbor This policy is based on Han et al. (1987) who aim on reducing travel times by selection of positions that are close to each other. There are many possibilities how this can be realized as *Nearest Neighbor* describes more a general idea than a predefined sequence. Therefore, we define three different rules that can serve as main routing policy:

1. **Simple**: This is choosing the nearest of the remaining fixed positions to be approached next. Physical restrictions within the cycle need to be considered, i.e., retrieval is only possible if the load handling device has free capacities.
2. **Reverse nearest neighbor (RNN)**: The four fixed positions are compared and arranged in such a way that the retrieval (storage) position closer to the I/O point is approached lastly (first). (Meller and Mungwattana 1997)
3. **Storage near retrieval / Retrieval near storage**: This rule is employed for storage or retrieval selection. Based on one fixed position, another

one is selected as near as possible from all potential positions, e.g. selection of a storage position close to a retrieval positions.

The first and the second rule belong to the case that **all positions are fixed** (regarding one cycle). The third requires a **free selection** of positions (either storage positions from all free positions or retrieval positions from the retrieval list).

Flip Flop *Flip Flop* is a policy that requires a dual load handling device. The quadruple command cycle is executed in the order SRSR. The first storage position is chosen randomly, while the second storage is performed into the position of the first retrieval. The first storage and second retrieval position are not defined furthermore. Consequently, if no additional rule is selected, they are chosen randomly.

Multiple Storage Facilitated by a dual load handling device, *Multiple Storage* strategies aim for a simultaneous execution of both storage jobs in adjacent positions. In the double deep environment, executing both storage jobs into a single empty storage lane represents another variant. For the strategies in Table 5.2, we always imply both variants of this concept. Important when using this concept is to find the appropriate storage positions. If no such positions are available, the cycle is executed randomly in the SSRR order.

Increase Tango The idea of the *Increase Tango* strategy is to limit the execution order to SSRR and to create more possibilities to perform a tango rearrangement. With this rule as a basis, the strategy can easily be combined with additional rules. In section 4.6.1, this way of execution has been evaluated. Therefore, in Table 5.2, *Increase Tango* is always combined with the idea of avoiding regular rearrangements and forcing tango rearrangements to further increase the amount of tango rearrangements.

Class based storage Two different occurrences of class based storage in Table 5.2 have to be distinguished. As **main routing policy**: According to

Schwarz et al. (1978) we define two strategies for application with classed based storage assignment (i.e., CB1 and CB2). They can not be transferred to random storage assignment and therefore belong to this separate main routing policy.

The second occurrence is discussed below in this subsection.

Additional rules

Nearest Neighbor When used as an additional rule, generally all three variants of the *Nearest Neighbor* policy can be applied to existing main routing policies.

No-Cost This additional rule is closely related to the *Shortest Leg* policy. In that case a position, either storage or retrieval, along the path of the S/R machine is selected. In both cases, free selection of positions is required.

Minimum Perimeter This additional rule is is a special case of the *Flip Flop* policy. The Minimum Perimeter heuristic, is presented by Keserla and Peters (1994). For the cycle, two retrieval positions that are both close to the first storage and close to each other are chosen. Afterwards, they are arranged in the travel-time minimal order.

Avoided or Forced Rearrangement This additional rule influences the occurrence of rearrangements by using the retrieval list: To avoid a rearrangement, a not-blocked retrieval request is deliberately chosen from the retrieval list. Applied vice verse, i.e., choosing a blocked retrieval request, forces a rearrangement. In that way, the *Increase Tango* policy is intensified.

Remaining elements

Selection of the single requests SP1 and SP2 are the first and second storage position, while RP1 and RP2 are the first and the second retrieval position approached within the cycle. The columns in this part of Table 5.2

indicate how the locations are selected. If no rule for the selection is defined and free selection (for storage positions or retrieval requests) is not applied, this is indicated by 'fixed'. If, in the case of free selection, no rule is applied, this is indicated by 'random'. Remember the Flip Flop example where the first storage position is chosen randomly.

Rearrangement rules For regular rearrangements, the nearest neighbor rule, as explained in section 4.2.2, is assumed. As this is an established and efficient way to perform rearrangements, we do not apply additional heuristics. Tango rearrangements are used for rearrangement in case both load handling devices are free.

Execution order This column indicates which execution orders are possible for the particular strategy. In many cases the execution order is determined by the routing policy, for example with multiple storage only SSRR is possible. If both execution orders are possible ('SSRR or SRSR'), we assume that the order is randomly chosen, i.e., $P(SSRR; SRSR) = (1/2; 1/2)$. However, in practice, a single execution order or a specific probability distribution of the execution orders could be selected.

Class based storage Class based storage also occurs as a separate element within **storage assignment** policies: Theoretically, all strategies can be applied with random or class-based storage assignment, but some need to be adjusted for application with class based storage. Most important when applying this concept is to assure that all articles are stored in the correct class. For strategies with fixed (predefined) storage positions, the correct class is selected automatically. For other strategies, we suggest the fowling adjustments in the main routing policies:

- *Shortest Leg* adjustments for strategies with a free selection of storage positions: Only available positions within the no-cost zone from the given class are permitted to choose. Positions available for storage that do not belong to the correct class must not be selected.

- *Nearest Neighbor* with a free selection of storage positions (i.e., storage near retrieval): Only available positions from the given class are permitted to choose. The nearest position may not belong to the correct class, such positions must not be selected.
- *Flip Flop* is accomplished in the following way: A retrieval request matching to the class of (at least) one storage request has to be selected to perform the Flip Flop operation (see Sarker, Mann and Leal Dos Santos (1994)). If no resorting of the retrieval list is possible, Flip Flop is only performed if one storage unit matches one of the classes of one retrieval request.
- *Multiple Storage* : If both storage units belong to the same class, perform the intended strategy. Otherwise store them separately according to their classes, which means multiple storage can not be applied.

Note that we do not propose how the classification is realized and the zones are designed.

5.3 Analytical Formulation for the Basic Strategies

The objective of this section is to analytically formulate the **main routing policies**. Since they represent recurring elements of all strategies, we thus provide the basis for travel time composition of many different strategies.

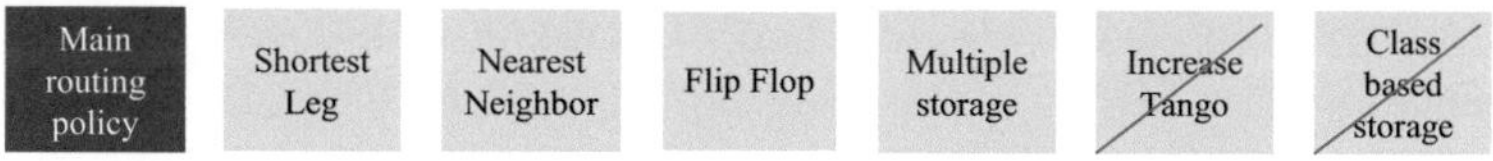

Figure 5.3: Main routing policies selected for analytical description

In fact, analytical formulations for the following four main routing policies are derived (see Figure 5.3):

- *Shortest Leg*
- *Nearest Neighbor*

- *Flip Flop*
- *Multiple Storage*

For Nearest Neighbor, the three different variants explained in section 5.2.2 are considered, i.e., two that operate under the assumption of fixed storage and retrieval positions and one for free selection of storage positions.

Not treated analytically are the main routing policies:

- *Increase Tango*
- *Class based storage*

The main idea of *Increase Tango* is consistent with the SSRR model and was already explained in subsection 4.3.2. The concept of *Class based storage* is highly dependent on the exact definition of the the classes and furthermore on the characteristics of the respective range of SKUs that are stored in the warehouse. Therefore, we do not address these two policies any further in this section.

Consistent with Chapter 4, we set up the models of this section based on the results of the general model and its assumptions. This means, unless stated otherwise, the assumptions of section 4.1.1 still hold. If not, additional or changed assumptions are stated in the respective subsections.

At the beginning of each section, we present **necessary adjustments** compared to the respective model of Chapter 4. Depending on the main routing policy, the underlying model differs with regard to the probability distribution of the execution order. This means, we make use of the universal approach represented in the general model from section 4.2. The presented examples from section 4.3 are again indicated with Basic model, SSRR model and SRSR model, their expected travel times from equations 4.51, 4.59 and 4.67 are applied respectively. In all cases, we derive **new expression** for each $E(QC)_{dd}$ by changing the $E(QC)$-part of the underlying model. If further **analytical components** are changed, such as mean load handling or rearrangement times, we explain them in the particular subsections.

Some of the strategies discussed in this section are based on the idea of reducing the cycle from four to three stops whereby one travel between distance is saved, for example *Flip Flop*. For those, the $E(QC)$-part, is adjusted in the following way (compare equation 4.3):

$$E(QC^{3S})^N = E(SC)^N + 2 \cdot E(TB) \tag{5.1}$$

This means, at the beginning of each of the following subsection, we present three key aspects:

1. The underlying model, respectively the probability distribution of the execution order.
2. The changed components from $E(QC)_{dd}$.
3. The new components needed.

In the following, SP1 refers to the first storage position and SP2 refers to the second storage position, while RP1 refers to the first retrieval position and RP2 refers to the second retrieval position within the cycle.

5.3.1 Nearest Neighbor

In the context of *Nearest Neighbor* we state three main rules in subsection 5.2.2. All of them are covered in the following. Their ID according to Table 5.2 is given in parentheses.

- **Simple**: Choosing the nearest location among a given number of fixed positions (NN1).
- **RNN**: Choosing the nearest locations seen from the I/O point (NN2).
- **Storage near retrieval**: Choosing the nearest location for storage from all open positions seen from a fixed retrieval position (NN5).

Table 5.3 gives an overview of all new components needed for analytical determination of the listed NN models.

The expected smallest one-way travel time from the I/O point to one of m open locations is derived by Meller and Mungwattana (1997). Comparing their results to Monte Carlo simulation, they can validate $E(SW_m)$. Using $m = 2$ in equation 3.61, we obtain the travel distance from the I/O point to the first stop of the cycle when applying NN1 and NN2.

Notation	Equ.	Explanation	Reference
$E(SW_m)$	3.61	Expected smallest one-way travel time from the I/O point to one of m randomly selected locations	Meller and Mungwattana (1997)
$E(TB_m)$	3.59	Expected smallest travel-between time between one point and m randomly selected locations	Han, McGinnis, Shieh and White (1987)
$E(TB_m^o)$	5.8	Expected o^{th} smallest travel-between time amongst one point and m randomly selected locations	

Table 5.3: Needed components for analytical determination of the listed NN models

$E(TB_m)$ is derived by Han et al. (1987) and validated by Monte Carlo simulation. Equation 3.59 is evaluated for $m = 2$ and $m = 3$ for application in NN1 and NN2. In NN5, we use the general formulation, because m is depending on the number of available storage positions that are available to store into.

$E(TB_m^o)$ is derived by Happes, Dörr and Furmans (2017). It is applied in NN5 to determine the rearrangements distance to the the second-nearest position, in case the nearest positions is already used for storing.

NN1: Simple Nearest Neighbor Rule

Underlying model	$P(SSRR; SRSR) = (\frac{1}{3}, \frac{2}{3})$
New components	• $E(SW_2)$ • $E(TB_2)$ • $E(TB_3)$

Table 5.4: Components for the derivation of the NN1 model

For the **simple** nearest neighbor model, the assumptions of section 4.1.1 are valid. Table 5.4 outlines the underlying model and the new components needed to define the model.

Figure 5.4 shows a comparison of the randomly performed QC with one possible NN1 cycle. The execution of the QC under the NN1 model is determined in the following way: Starting at the I/O point, the storage position which is located closer to the I/O point is approached first (SP1 or SP2). The selection is equally likely, hence both, SP1 or SP2, have a chance of $\frac{1}{2}$ being located closer to the I/O point. Without loss of generality, assume SP1 is chosen. Afterwards, the closest of the remaining three positions is selected, i.e., SP2, RP1 or RP2. Since the positions are randomly selected from the storage rack, the chance to select a retrieval position next is $\frac{2}{3}$. Therefore, the probability distribution of the execution order is $P(SSRR; SRSR) = (\frac{1}{3}; \frac{2}{3})$.

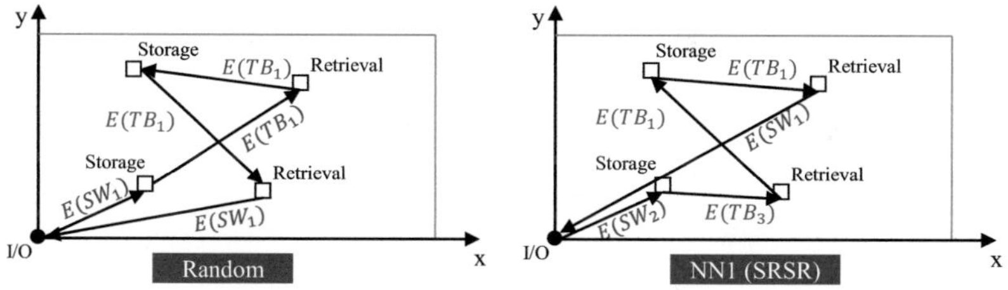

Figure 5.4: Example of a NN1 cycle (in SRSR order) with the path-depending components of the cycle

In order to select the third stop, we need to distinguish between SSRR and the SRSR. Within the SRSR order, i.e., RP1 or RP2 are the second stop, no further choice of positions is possible and SP2 is stored next. If, on the contrary, SP2 is selected for the second stop, resulting in the SSRR order, the closest among both retrieval positions, RP1 or RP2, is selected. In both cases, the fourth stop is the reaming retrieval location. With the last trip, the S/R machine returns to the I/O point.

The determination of the probabilities in every step and both orders can be illustrated using a tree diagram that can be found in the Appendix (Figure A.2). We do not present a model for the probability distribution of the execution order of $P(SSRR; SRSR) = (\frac{1}{3}; \frac{2}{3})$ so far. Therefore, a calculation example of that model can be found in the Appendix A.

For the determination of the NN1 model, the path-depending travel time components are changed. Depending on the execution order the new components are applied according to the procedure of the cycle described above. How the individual components are replaced compared to the underlying model is illustrated in Figure 5.5.

	General model	NN1 model	
		SSRR	SRSR
1	$E(SW_1)$	$E(SW_2)$	$E(SW_2)$
2	$E(TB_1)$	$E(TB_3)$	$E(TB_3)$
3	$E(TB_1)$	$E(TB_2)$	$E(TB_1)$
4	$E(TB_1)$	$E(TB_1)$	$E(TB_1)$
5	$E(SW_1)$	$E(SW_1)$	$E(SW_1)$

Figure 5.5: Cycle time components of the NN1 model compared to the basic model

We obtain the following results by solving equations 3.61 and 3.59 for $m = 2$ and $m = 3$.

$$E(SW_2) = \frac{2}{15}b^3 + \frac{2}{3} \tag{5.2}$$

$$E(TB_2) = -\frac{11}{630}b^5 + \frac{5}{42}b^4 - \frac{31}{105}b^3 + \frac{1}{3}b^2 + \frac{1}{5} \tag{5.3}$$

$$E(TB_3) = -\frac{107}{10010}b^7 + \frac{31}{330}b^6 - \frac{793}{2310}b^5 + \frac{143}{210}b^4 - \frac{11}{14}b^3 + \frac{1}{2}b^2 + \frac{1}{7} \tag{5.4}$$

Considering the probability distribution of the execution order and the new components, the $E(QC)^N$ term of the quadruple command cycle from the $(\frac{1}{3},\frac{2}{3})$ model is changed as follows:

$$\begin{aligned} E(QC)^N = \\ E(SW_2) + \tfrac{E(TB_3)+E(TB_2)+E(TB_1)}{3} + \tfrac{2}{3}[E(TB_3) + 2E(TB_1)] \\ +E(SW_1) \end{aligned} \tag{5.5}$$

Inserting 5.5 into the equation A.13 presented in the Appendix gives the mean travel time for NN1.

NN2: Reverse Nearest Neighbor Rule

All assumptions of section 4.1.1 still hold for NN2. Again, the execution order is not chosen randomly, but determined according to the RNN heuristic.

Underlying model	Basic model; $P(SSRR;SRSR) = (\frac{1}{2},\frac{1}{2})$
New components	• $E(SW_2)$

Table 5.5: Components for the derivation of the NN2 model

The order of the (predefined) positions within the QC is determined as follows: First, among both storage positions and both retrieval positions, those positions located closer to the I/O point are selected. Without loss of generality, assume SP1 and RP1 are chosen. SP1 is scheduled to be the first stop within the cycle, RP2 is the last stop in the cycle. The travel time to both positions, SP1 and RP1, is described by the smallest one-way travel time from the I/O point to one of two randomly selected positions, i.e., $E(SW_2)$. The position for the second stop is randomly selected and results in the appropriate execution order (SSRR or SRSR). SP2 and RP2 are randomly chosen and thus have the same probability of being approached next. Therefore, the probability distribution of the execution order is $P(SSRR;SRSR) = (\frac{1}{2},\frac{1}{2})$ and the analytical model for NN2 is based on the basic model (see 4.51). Note that this is different to NN1, because in the second step of NN1 we chose from three instead of two positions. Figure 5.6 illustrates the travel time components of the NN2 strategy in comparison to the the random executed cycle.

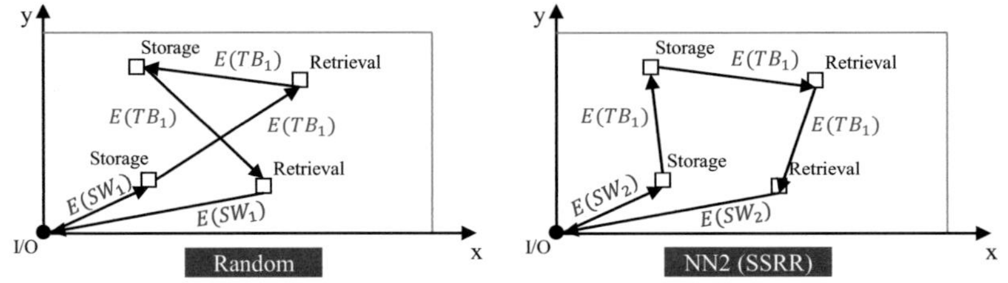

Figure 5.6: Example of a NN2 cycle with the path-depending components of the cycle

Again, a decision tree illustrating the definition of the cycle can be found in the Appendix (Figure A.3).

Substituting the underlying model by the mentioned components for the first and fifth travel movement of the cycle, $E(QC)^N$ can be changed accordingly:

$$E(QC)^N = 2E(SW_2) + 3E(TB_1) \tag{5.6}$$

Inserting 5.6 into the model given in equation 4.51 results in the mean travel time for NN2.

Note that only a small adaption is needed to model NN3 from Table 5.2. For NN3, the second stop is chosen from the two remaining positions, SP2 and RP2, in such a way that the smallest travel time is generated. The time between a given position and two randomly selected positions is given by $E(TB_2)$. Therefore not only the first and fifth travel movement change in comparison to the underlying model, but also the second one. Thus $E(QC)^N$ is changed in the following way:

$$E(QC)^N = 2E(SW_2) + E(TB_2) + 2E(TB_1) \tag{5.7}$$

Again, inserting 5.7 into equation 4.51 results in the mean travel time for NN3.

NN5: Storage near Retrieval Rule

For the nearest neighbor policy of NN5, we allow to choose storage positions freely from all open locations instead using predefined storage locations. The following assumptions are changed compared to section 4.1.1.

- Storage operations are performed according to a nearest neighbor policy, i.e., the nearest available storage position of each retrieval position is assigned to the preceding storage job.
- The execution order therefore is determined to be SRSR.
- Effects at the edge of the rack, such as a smaller probability of finding an available storage location, are ignored.

In Table 5.6, an overview of the needed components is given.

Underlying model	SRSR model; $P(SSRR; SRSR) = (0, 1)$
New components	• $E(TB_m^o)$ • $E(TB_m)$

Table 5.6: Components for the derivation of the NN5 model

Because of the execution order, the underlying model is the SRSR model. In contrast to NN1 and NN3, storage positions are not predefined. The two available storage positions nearest to the fixed retrieval positions are selected for storing. In case a rearrangement is required, the second nearest position is used for rearrangements.

The expected o^{th} smallest travel-between time between one point and m randomly selected locations is defined by Happes et al. (2017):

$$E(TB_m^o) = \int_0^1 \zeta \frac{m!}{(o-1)!(m-o)!} q(\zeta) Q(\zeta)^{(o-1)} (1 - Q(\zeta))^{(m-o)} d\zeta \qquad (5.8)$$

$Q(\zeta)$ and $q(\zeta)$ are the density function and the distribution function of the travel between distances (see equations 3.48 and 3.49) in a normalized ($1\,x\,b$) rack model. More details about the derivation of equation 5.8 can be found in the Appendix A. Figure 5.7 illustrates the different path-depending travel components of a NN5 cycle.

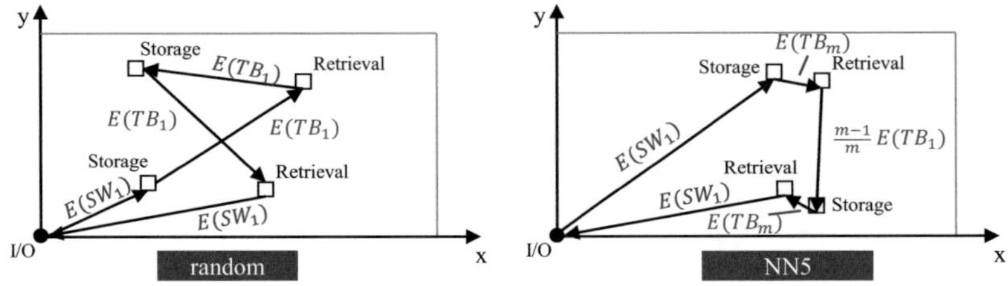

Figure 5.7: Example of a NN5 cycle with the path-depending components of the cycle

A cycle according to NN5 is defined in the following way: Among the two predefined retrieval positions, the specific order is chosen randomly. Without loss of generality, assume RP1 to be the first and RP2 to be the second retrieval position. To determine the first stop of the cycle, we choose the storage position closest to RP1, which is SP1. The travel time between SP1 and RP1 is $E(TB_m)$ with m being the number of all open locations (we use pR from equation 4.34). The third stop, SP2, is defined in the same way, i.e., the closest of all potential storage locations in relation to RP2. Again, the travel time between SP2 and RP2 is $E(TB_m)$. Furthermore, the travel time between RP1 and SP2 needs to be determined. Here, we have to distinguish two different cases:

Case 1: $RP1 \neq SP2$

For randomly chosen retrieval positions and m open locations in the storage space, $m-1$ positions are different from that one just being released trough RP1. Note that from the transition probabilities of the SRSR model follows that after completion of SP1 and RP1 the mean number of open position remains m (Happes, Dörr and Furmans 2017, p. 15). Therefore, with a probability of $\frac{m-1}{m}$ the location to store SP2 does not coincide with RP1. For this case the travel between time is given by $E(TB_1)$.

Case 2: $RP1 = SP2$

Consequently, in one out of m cases the position of SP2 is equal to RP1. No movement of the S/R machine is required and the travel between distance is zero. The probability for this case is $\frac{1}{m}$.

In total, the travel time between RP1 and SP2 is $\frac{m-1}{m}E(TB_1)$. The first and the last travel distance of any cycle correspond to $E(SW_1)$. The path-depending travel time therefore adds up to:

$$E(QC)^N = 2E(SW_1) + \left(2E(TB_m) + \frac{m-1}{m}E(TB_1)\right) \tag{5.9}$$

	General model	NN5 model: Rearr. into E	NN5 model: Rearr. into H
I/O to SP1	$E(SW_1)$	$E(SW_1)$	
SP1 to RP1	$E(TB_1)$	$E(TB_m)$	
RP1 to SP2	$E(TB_1)$	$\frac{m-1}{m}E(TB_1)$	
SP2 to RP2	$E(TB_1)$	$E(TB_m)$	
RP2 to I/O	$E(SW_1)$	$E(SW_1)$	
Rearr. distance E(RD)	$(\frac{7}{15})^{1-\frac{pR}{F*}}\frac{1}{\sqrt{pR}}$	$E(TB_{pR})$	$E(TB^2_{pR})$

Figure 5.8: Cycle time components of the NN5 model compared to the basic model

To specify the travel time of NN5, also rearrangements are considered. If a rearrangement is needed for one of the retrievals, the nearest neighbor policy still applies. The closest open position may not exist any more due to a preceding nearest neighbor storage operation. This happens if the storage lane chosen for storage was in the half-filled state. In this case, potential rearrangements have to be performed into the second nearest position, which is analytically expressed by $E(TB^2_m)$. In case the storage lane was in state E, a rearrangement into the same lane is still possible. Therefore, we have to distinguish between storage operations performed into empty and half-filled storage lanes when determining the rearrangement travel time. We use the probabilities for the different storage events based on Lippolt (2003). The probabilities of storing into an empty storage lane and a half-filled storage lane, respectively, are:

$$P(S1)_{SRSR} = \frac{1}{2z+1} \tag{5.10}$$

$$P(S2)_{SRSR} = \frac{2z}{2z+1} \tag{5.11}$$

Accordingly, the expected rearrangement distance is adjusted:

$$E(RD) = P(S1)_{SRSR} \cdot E(TB_{pR}) + P(S2)_{SRSR} \cdot E(TB^2_{pR}) \tag{5.12}$$

When inserting this into equation 4.42, we obtain the rearrangement time for the NN5 model. Combining the new components and substituting the different parts into equation 4.67 gives the new requested mean travel time.

Note that using the SRSR model is only an approximation of the underlying storage lane allocation. Due to the case differentiation for the travel between RP1 and SP2, in $\frac{1}{m}$ cases the underlying model is different. Using Monte Carlo simulation with 10^6 samples, it can be shown that the deviation of equation 5.9 to the exact value is less than 0.1%.

Strategies with nearest neighbor selection based on retrieval sequencing

Both equations, 3.59 and 5.8, represent an important basis for the derivation of many other strategies using a nearest neighbor selection. Whenever a unit for retrieval should be selected near to a given position, $E(TB_m)$ applies. While for NN5, all available positions for storage represent m, for retrieval selection, the number of applicable positions, i.e., stored units, is depending on a greater number of individual parameters, such as sequencing possibilities and number of units per SKU.

5.3.2 Flip Flop (FF1)

Based on the description of Flip Flop in section 5.2, we consider a randomly performed cycle including a Flip Flop movement. The following assumptions are changed compared to section 4.1.1.

- The execution order is determined to be SRSR.
- The second storage unit is stored into the first retrieval position.

Table 5.7 shows the underlying model and analytical components.

Underlying model	SRSR model; $P(SSRR;SRSR) = (0,1)$
Changed components	$E(t^S_{LHD})$

Table 5.7: Components for the derivation of the FF1 model

The model we build upon is the SRSR model, because the execution order SRSR is required for Flip Flop. Figure 5.9 illustrates the analytical elements of the Flip Flop cycle compared to random execution.

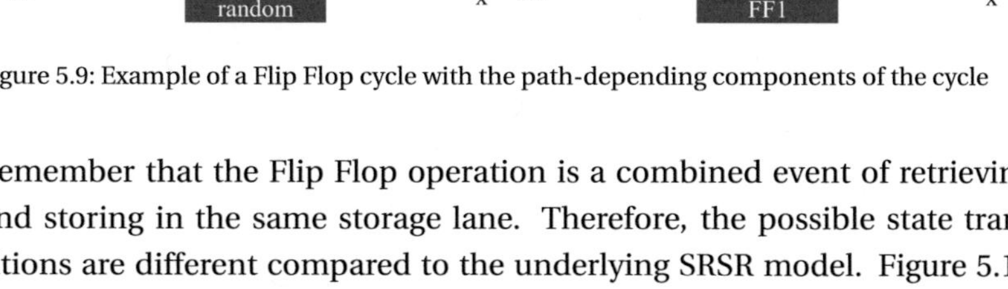

Figure 5.9: Example of a Flip Flop cycle with the path-depending components of the cycle

Remember that the Flip Flop operation is a combined event of retrieving and storing in the same storage lane. Therefore, the possible state transitions are different compared to the underlying SRSR model. Figure 5.10 shows all possible transitions of a storage lane during a Flip Flop operation. As retrieval operations are part of a Flip Flop, also rearrangements can occur. There are four possible storage lane transitions when performing a Flip

Flop. The storage unit is placed into the released position of the retrieval unit. Rearrangements can be performed according to the options from the general model (i.e., into an empty or half-filled storage lane). The transitions during the remaining operations, i.e., the first storage and the second retrieval, are not shown as they do not change compared to the underlying model.

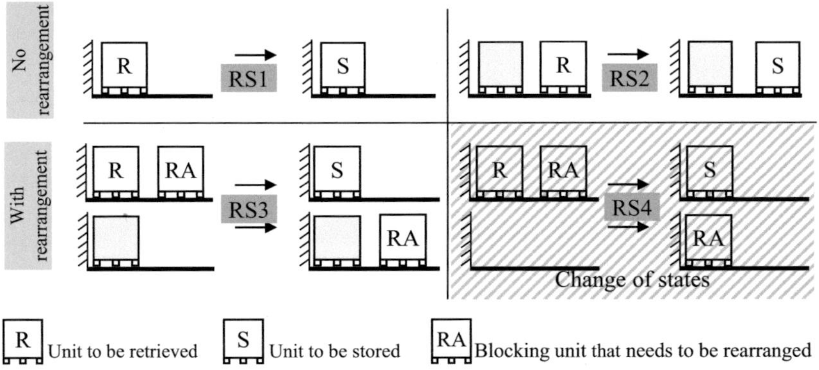

Figure 5.10: Possible state transitions for a Flip Flop operation

RS1 - RS3 in Figure 5.10 exhibit the same allocation of storage lanes, before and after the Flip Flop. This is different for RS4 (shaded). Here the state of the storage lane allocations is changed, because the blocking unit is stored into an empty storage lane and the storage unit is stored into the now empty storage lane. If we allow transition RS4, storage transitions and consequently the stationary allocation do not correspond to the underlying SRSR model. To overcome this obstacle we assume the following: If during a Flip Flop operation a rearrangement is necessary and an empty storage lane is chosen as rearrangement location, also the storage unit is stored into the rearrangement lane. Figure 5.11 shows this adapted version of RS4.

If the transition RS4 is changed like this, the state transitions are equivalent to those of the SRSR model and we can apply it as the underlying model.

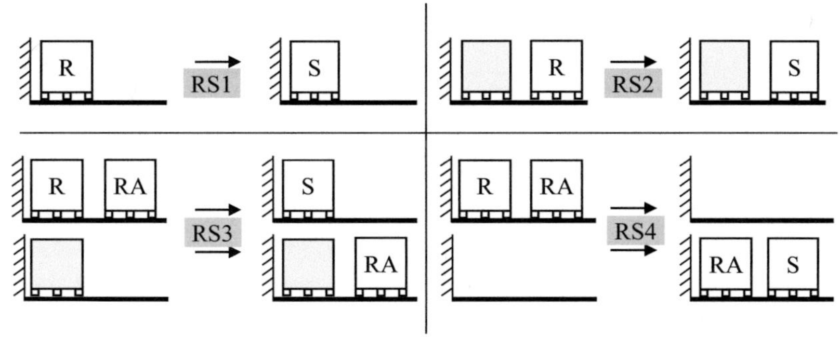

Figure 5.11: Possible state transitions for the Flip Flop movements with adjustment for RS4

Next, consider the mean load handling times for storage operations. We distinguish between the first and second storage operation, because only the second one is affected by Flip Flop, while the first one is a 'regular' storage. For the second storage operation, the load handling times depend on the kind of the Flip Flop movement, i.e., which of the four transitions RS1-RS4 is performed. For the events RS1 and RS3, storage operations are performed into the rear position. They originate from the combination of transitions S1 with R1 and R3, respectively (see Figure 4.5 and Figure 4.6). For RS2 and RS4 the storage is performed to the front position of a storage lane. RS2 follows from transition R2, while RS4 results from transition R4. The mean load handling time of a storage operation that occurs during the Flip Flop is:

$$E(t_{LHD}^{S^{FF}}) = \frac{1}{2} \cdot (t_{LHD,f} \cdot (1 + P(R2) + P(R4)) + t_{LHD,r} \cdot (P(R1) + P(R3))) \tag{5.13}$$

For the first storage, the load handling times of the underlying model (see eqaution 4.65) are relevant. To determine an overall mean load handling time for both storage operations, we average the load handling times of the first and second storage operation. This results in:

$$E(t^S_{LHD})_{FF} = \underbrace{\frac{1}{2} \cdot E(t^S_{LHD})_{SRSR}}_{First\,storage} + \underbrace{\frac{1}{2} \cdot E(t^{S^{FF}}_{LHD})}_{Second\,storage} \tag{5.14}$$

During each Flip Flop, there is a deterministic sidewards movement of the S/R machine for repositioning of the LHDs. Therefore, t_d needs to be incorporated into the formula.

Adjusting dwell times and the number of acceleration phases, the mean travel time for a Flip Flop cycle is:

$$\begin{aligned}
E(QC_{dd})_{FF} = {} & t_0{}^3 + t_d + 2 \cdot (\tfrac{v_x}{a_x} + \tfrac{v_y}{a_y}) + E(QC^{3S})^N \cdot \tfrac{L}{v_x} \\
& +2 \cdot P_{Rearrange} \cdot (\\
& \quad t_{0,Rearr.} + (\tfrac{v_x}{a_x} + \tfrac{v_y}{a_y}) + 2 \cdot E(t^S_{LHD})_{FF} + 2 \cdot E(RD) \cdot \tfrac{L}{v_x}) \\
& +4 \cdot E(t^S_{LHD})_{FF} + 4 \cdot E(t^R_{LHD}) - 2 \cdot t_{LHD,f}
\end{aligned} \tag{5.15}$$

Note that equations 4.64, 4.63, 4.56, 4.65 and 4.66 apply here.

5.3.3 Shortest Leg (SL1)

In this subsection, the modeling of Shortest Leg corresponding to strategy SL1 is addressed. The following assumptions in addition to section 4.1.1 apply here.

- The execution order is determined to be SSRR.
- The second storage position is an open location within the no-cost zone between SP1 and RP1. If no such position is available, the cycle is performed randomly according to the underlying model.
- The dimension of each storage lane has the same proportion as the rack.

3 We assume $5 \cdot t_{mast}$ exactly like for the random execution. It may by possible that t_0 consists of $4 \cdot t_{mast}$, see subsection 4.2.3

Table 5.8 summarizes all necessary components for the model.

Underlying model	SSRR model: $P(SSRR; SRSR) = (1,0)$
Changed components	*none*
New components	$P(SL)$

Table 5.8: Components for the derivation of the SL1 model

The aim of the shortest leg idea is to save one travel between distance, which can be realized if there is an empty position within the no-cost zone between SP1 and RP1. The travel time components of the cycle in comparison to a random executed cycle are shown in Figure 5.12.

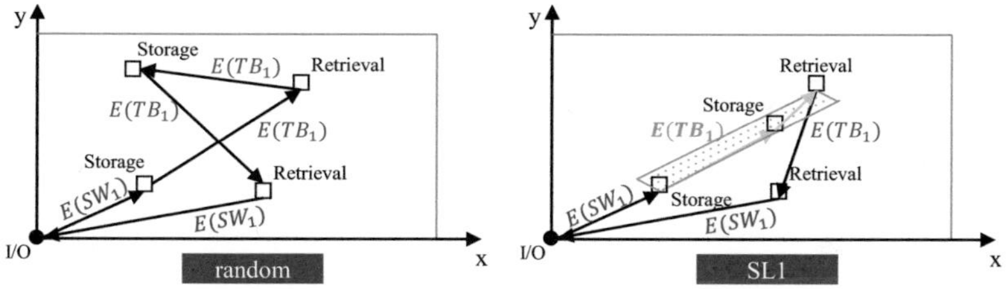

Figure 5.12: Example of a SL1 cycle with the path-depending components of the cycle

No new analytical expressions are needed for the path-depending travel movements. Instead, we need to define the **probability** that there is an **empty position within the no-cost zone**, P(SL). Therefore, the size of the no-cost zone and the storage allocation within the zone are required.

Consider any mean travel between two randomly chosen positions, (x_1, y_1) and (x_2, y_2), within a normalized, dimensionless rack as shown in Figure 5.13. From section 3.2.1, we know the expected distance between them is 7/15. One of the two distances, in x- or y-direction, is the greater one and therefore determines the travel time. Without loss of generality, let $|x_1 - x_2| \geq |y_1 - y_2|$, then $|x_1 - x_2|$ is the normalized travel time. In the direction with the smaller distance, there is a range for additional movement of the S/R machine. This range is $|x_1 - x_2| - |y_1 - y_2|$ as illustrated in Figure 5.13.

Within this range, the S/R machine can move to a greater area without loss of time. Figure 5.13 illustrates that from the angle α, that is formed by the horizontal axis and the connection between the two positions, the range is determined. (Note that for $|x_1 - x_2| \le |y_1 - y_2|$, the angle is formed by the vertical and the connecting line).

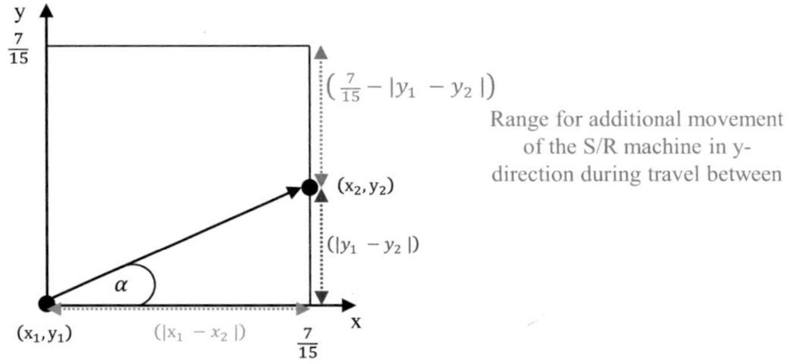

Figure 5.13: Travel between in a normalized, dimensionless rectangle showing the range in the shorter direction and the angle α.

To derive the mean size of the no-cost zone, the average value of α is needed with $\alpha \in [0, 45°]$. With the mean value theorem of integral calculations, the expected distance in the smaller direction and thus the average value of α can be calculated. The result is $\alpha = 23.81049°$ (Brunk 2016). Knowing the average value of α, the range for additional movement is $\frac{7}{15} - \frac{7}{15}\tan(\alpha)$. This is used to define the exact dimensions of the mean no-cost zone as shown in Figure 5.14.

To calculate the surface area of the no-cost zone ($A_{no-cost}$) its side lengths are needed (within the normalized $(1x1)$-rack). The line between SP1 and P2 is the hypotenuse of the right, isosceles triangle with leg length of 0.13037. Using the theorem of Pythagoras, the length is determined as 0.184371. The line between P2 and RP1 is the hypotenuse of the right, isosceles triangle with length of 0.336297. Using the theorem of Pythagoras, the length is determined as 0.475595.

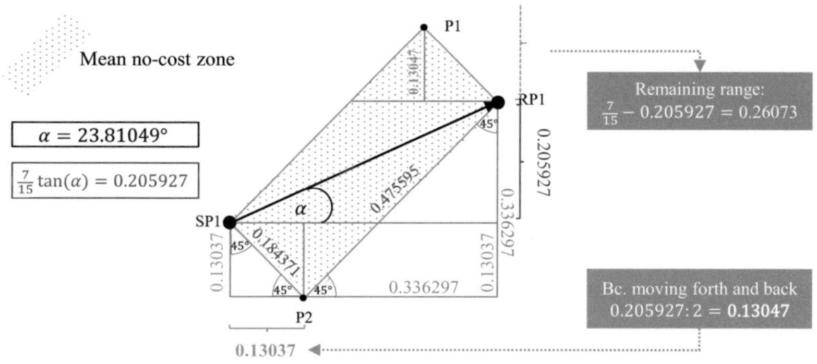

Figure 5.14: Dimensions of the mean no-cost zone

The mean size of the no-cost zone can be calculated according to its dimension, which is:

$$A_{no-cost} = 0.184371 \cdot 0.475595 = 0.087686 \tag{5.16}$$

Given the total number of storage lanes of the rack (l), the mean number of storage lanes within the no-cost zone can be determined:

$$Nl_{no-cost} = A_{no-cost} \cdot l = 0.087686 \cdot l \tag{5.17}$$

Knowing the mean total number of positions in the no-cost zone as well as the mean state probabilities of the storage lanes, we can determine the expected number of available storage positions within the no-cost zone and thus the probability of finding at least one position that is available for storage. We can use the complementary probability, i.e., that no position is available within the no-cost zone. With $P(F)$ from the SSRR model (equation 4.54), the probability that all positions within the no-cost zone are fully occupied is $P(F)^{Nl_{no-cost}}$. Consequently, the probability that performing the shortest leg cycle is possible is:

$$P(SL) = 1 - P(F)^{Nl_{no-cost}} \tag{5.18}$$

If the shortest leg cycle is performed, $E(QC^{3S})^N$ applies for the path-depending travel time. Dwell times and acceleration parts remain unchanged as the number of stops is not reduced. Using equation 4.59 in case no shortest leg cycle is possible, the mean travel time of the main routing policy Shortest Leg is given by:

$$\begin{aligned}E(QC_{dd})_{SL} = t_0 + \frac{5}{2} \cdot \left(\frac{v_x}{a_x} + \frac{v_y}{a_y}\right) \\ + (1 - P(SL)) \cdot E(QC)^N \cdot \frac{L}{v_x} + P(SL) \cdot E(QC^{3S})^N \cdot \frac{L}{v_x} \\ + 1 \cdot P_{Rearrange} \cdot E(RC) + 1 \cdot P_{Rearrange} \cdot t_{Tango} \\ + 4 \cdot E(t^S_{LHD}) + 4 \cdot E(t^R_{LHD}) - 2 \cdot t_{LHD,f}\end{aligned} \tag{5.19}$$

Note that equations 4.40, 4.42, 4.55, 4.56, 4.57 and 4.58 apply here.

It is important to mention that the derivation of the area of the no-cost zone is determined based on the continuous rack model. Transferring the result to a discrete rack with a finite number of storage lanes (l), causes inaccuracies due to discretization and rounding errors.

5.3.4 Multiple Storage (MS1)

In modeling Multiple Storage we distinguish between two variants of this strategy: Storing in empty positions next to each other (simultaneously) as well as storing both storage units into one empty storage lane (successively).

Simultaneous Multiple Storage (MS Sim)

For simultaneous multiple storage, both storage operations are performed simultaneously into two storage lanes next to each other. Thus, one travel between distance is saved compared to the basic model. Whether the cycle can be executed in the described way depends on the allocation of the storage lanes, i.e., whether adequate storage lanes are available. Therefore, the probability of their occurrence needs to be determined. Next, the following assumptions are made

- The first storage position is chosen randomly, the possibility of performing a multiple storage is evaluated depending on that position. In this way, the state probabilities of the storage lanes from Chapter 4 are still valid and can be used to derive the probability of occurrence. If multiple storage is not possible, the cycle is randomly performed according to the SSRR model.
- Availability of adequate storage capacity is the only limiting factor of multiple storage. No technical or physical restrictions are considered.
- Effects at the edge of the rack, such as a smaller probability of performing multiple storage, are ignored.

Underlying model	SSRR model; $P(SSRR;SRSR) = (1,0)$
Changed components	$E(t_{LHD}^{S})$ (when applying multiple storing)
New components	$P(MS.Sim)$

Table 5.9: Components for the derivation of the MS1 (simultaneous) model

The travel time components of the cycle in comparison to a randomly executed cycle are shown in Figure 5.15.

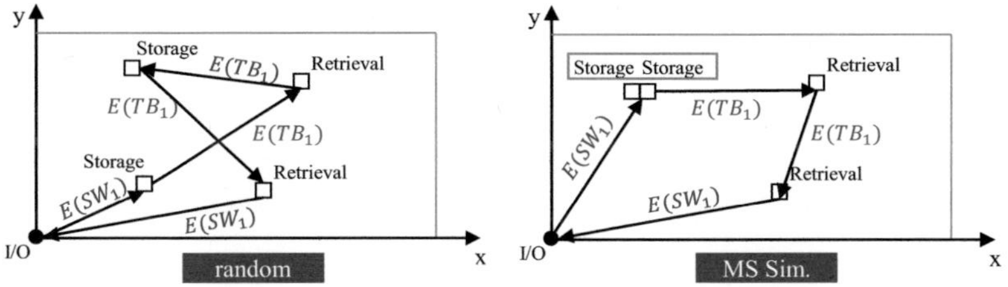

Figure 5.15: Example of a MS1 cycle with the path-depending components of the cycle

First, we define the **probability** of being able to **perform a simultaneous multiple storage**, $P(MS.Sim)$. Figure 5.16 shows the different states that two adjacent storage lanes can have. Multiple storage is always possible if both lanes have at least one empty storage position. The probability of performing multiple storage is given as the probability that two random storage lanes are obtaining state E or state H.

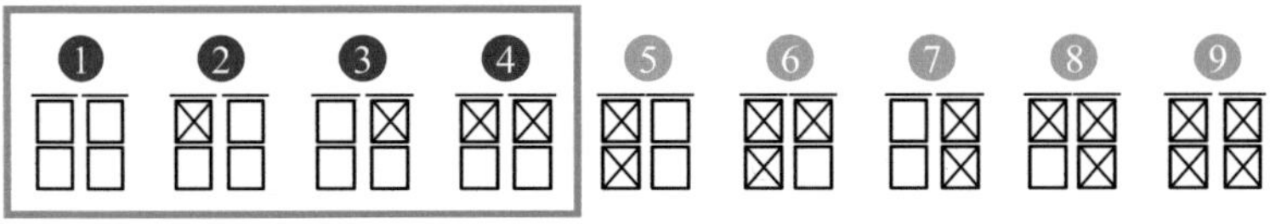

Figure 5.16: Different storage lane allocation hat two adjacent storage lanes can exhibit

With the state probabilities of the SSRR model (see equations 4.52 and 4.53), this can be defined as follows:

$$P(MS.Sim) = (P(E) + P(H))^2 \tag{5.20}$$

Apart from the path-depending travel times, dwell times and amounts for acceleration and deceleration change. The travel time when performing multiple storage operations without rearrangements and load handling is:

$$10t_{dead} + 4t_{mast} + 2(\frac{v_x}{a_x} + \frac{v_y}{a_y}) + E(QC^{3S})^N \tag{5.21}$$

Next, we consider the load handling times. Due to the simultaneous storage operation, the mean load handling time for storage operations changes compared to the underlying model. When storing into two adjacent storage lanes, the longer access time of the two is determining the time needed for the storage operation as a whole. If both storage lanes are in state H (case 4 in Figure 5.16), the time needed is $t_{LHD,f}$. Otherwise (cases 1 - 3 in Figure 5.16), the time needed is $t_{LHD,r}$. Therefore, the probability that the access time of multiple storage equals $t_{LHD,f}$ is:

$$P(LHD_f) = \frac{P(H)^2}{(P(E) + P(H))^2} \tag{5.22}$$

In the travel time model, equation 4.57 is changed based on the derivation of equation 4.37 in the following way:

$$E(t^S_{LHD})_{MS.Sim} = \frac{1}{2}(t_{LHD,f} \cdot (1 + P(LHD_f)) + t_{LHD,r} \cdot (1 - P(LHD_f))) \quad (5.23)$$

Based on equation 4.59, we can formulate the mean travel time for the simultaneous version of MS1 by:

$$\begin{aligned} E(QC_{dd})_{MS.Sim} &= (1 - P(MS.Sim)) \cdot E(QC)_{dd} \\ &+ P(MS.Sim) \cdot [\, 10t_{dead} + 4t_{mast} + 2(\frac{v_x}{a_x} + \frac{v_y}{a_y}) + E(QC^{3S})^N \cdot \frac{L}{v_x} \\ &+ 1 \cdot P_{Rearrange} \cdot E(RC) + 1 \cdot P_{Rearrange} \cdot t_{Tango} \\ &+ 4 \cdot E(t^R_{LHD}) - t_{LHD,f} + 2E(t^S_{LHD})_{MS.Sim} \,] \end{aligned} \quad (5.24)$$

Note that equations 4.42, 4.55, 4.56, 4.58 and 4.59 apply here.

The opportunity of performing multiple storage operations is depending on a random variable with this approach for $P(MS.Sim)$. However when implementing this operating strategy in practice, it is much more likely to be used whenever possible, i.e., as long as a pair of adequate storage lanes is available. For this reason, we adapt the occurrence probability from equation 5.20 based on an idea of Seemüller (2006, p. 151). He suggests to incorporate the generating probability from retrieval operations, which is the probability that a pair of storage lanes appropriate for multiple storage emerges by retrieval operations. Two conditions have to be fulfilled:

1. A storage lane in state F is chosen for retrieval which corresponds to the retrieval operations R2, R4 and R5 (Retrieving from state H would not generate an additional pair).
2. At least one neighboring storage lane of the considered storage lane in state F offers the possibility to store into (i.e., is in state E or H).

The generating probability for a new pair of storage lanes that offers the possibility for multiple storage is:

$$P(GP.Sim) = 2 \cdot \frac{4P(E)P(F) + 3P(H)P(F)}{2 \cdot (P(H) + 2P(F)) \cdot (P(E) + P(H))} \cdot (1 - P(F)^2) \qquad (5.25)$$

As long as the sum of the former probability of occurrence and the generating probability is greater or equal one, multiple storage is possible. The new probability for occurrence of multiple storage is:

$$\hat{P}(MS.Sim) = min\{1, (P(GP.Sim) + (P(E) + P(H))^2)\} \qquad (5.26)$$

Inserting this into equation 5.24 presents an adjusted travel time model that allows for a more practical application of MS1, enforcing the usage of multiple storage whenever possible.

The travel time model of equation 5.24 is approximated because of the storage lane allocation. We assume the storage lane allocation of neighboring storage lanes to be independent of each other. However, this is not guaranteed for simultaneous multiple storage in neighboring lanes.

Successive Multiple Storage (MS Suc)

For successive multiple storage, both storage operations are performed into one empty storage lane. Thus, one travel between distance is saved compared to the basic model. However, a small movement of the S/R machine that corresponds to t_d is needed instead. Whether the strategy can be executed is depending on the availability of empty storage lanes. Therefore, the probability of occurrence needs to be determined. Precisely, the following assumption is made:

- As long as empty storage lanes are available, one of them is chosen and a cycle with successive multiple storage is performed.

Table 5.9 summarizes the information about the underlying model, changed as well as new components.

Underlying model	SSRR model; $P(SSRR;SRSR) = (1,0)$
Changed components	$E(t^S_{LHD})$ (when applying multiple storing)
New components	$P(MS.Suc)$

Table 5.10: Components for the derivation of the MS1 (successive) model

The travel time components of the cycle in comparison with a randomly executed cycle are shown in Figure 5.17.

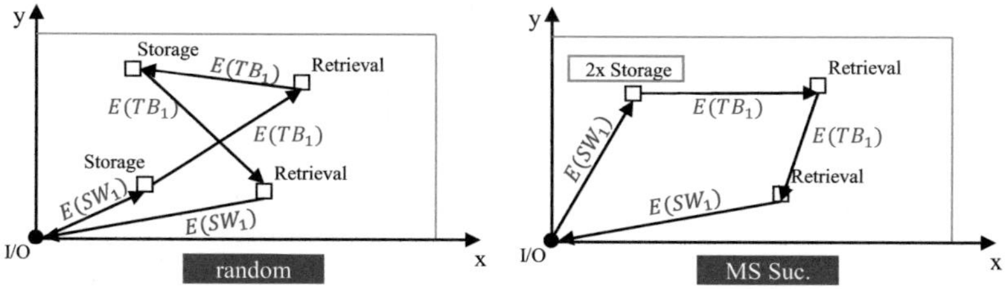

Figure 5.17: Example of a MS1 cycle with the path-depending components of the cycle.

The probability that successive multiple storage is possible is derived analogously to the case of simultaneous multiple storage. The random selection of the storage positions is combined with the generating probability of empty storage lanes to mirror realistic implementation.

Again, the first part is determined by the storage lane chosen for the first storage operation. The probability that an empty storage lane is chosen for storage is $\frac{P(E)}{P(E)+P(H)}$. The generating probability is the likelihood that an empty storage lane emerges by retrievals. Two retrieval operations, R1 and R3, produce empty storage lanes. Using the state probabilities of the storage lanes of the SSRR model (i.e., equations 4.52, 4.53 and 4.54), we obtain the generating probability for successive multiple storage:

$$P(GP.Suc) = \frac{2P(H) \cdot (P(E) + P(H)) + P(H)P(F)}{(P(H) + 2P(F)) \cdot (P(E) + P(H))} \tag{5.27}$$

The preceding factor 2 is necessary to allow for both retrieval operations within the quadruple command cycle. The probability of performing successive multiple storage is limited to one and is given by:

$$\hat{P}(MS.Suc) = min\{1, (\frac{P(H)}{P(E)+P(H)} + P(GP.Suc))\} \tag{5.28}$$

Similar to the simultaneous case, the load handling times are considered. For successive multiple storage, the storage process during multiple storage is a deterministic operation, that consists of the three following steps.

1. Storing to the rear position ($t_{LHD,r}$).
2. Moving the S/R machine sideways to position the other load handling device in front of the storage lane (t_d).
3. Storing to the front position ($t_{LHD,f}$).

In the travel time model, equation 4.57 therefore needs to be changed based on the derivation of equation 4.37 in the following way:

$$E(t^S_{LHD})_{MS.Suc} = \frac{1}{2}(t_{LHD,f} + t_{LHD,r} + t_d + t_{LHD,f}) \tag{5.29}$$

Based on equation 4.59 we can formulate the mean travel time for the successive version of MS1 by:

$$\begin{aligned} E(QC_{dd})_{MS.Suc} &= t_0{}^4 + \frac{5}{2}\cdot(\frac{v_x}{a_x} + \frac{v_y}{a_y}) \\ &+ (1 - P(MS.Suc))\cdot(E(QC)^N \cdot \frac{L}{v_x} + 4\cdot E(t^S_{LHD}) - t_{LHD,f}) \\ &+ P(MS.Suc)\cdot(E(QC^{3S})^N \cdot \frac{L}{v_x} + 2E(t^S_{LHD})_{MS.Suc}) \\ &+ 4\cdot E(t^R_{LHD}) - t_{LHD,f} + P_{Rearrange}\cdot E(RC) + P_{Rearrange}\cdot t_{Tango} \end{aligned} \tag{5.30}$$

Note that equations 4.42, 4.55, 4.56, 4.57, 4.58 and 4.3 apply here.

[4] We assume $5 \cdot t_{mast}$, both for the random cycle and the cycle with multiple storage. It may possible that t_0 consists of $4 \cdot t_{mast}$, see subsection 4.2.3

The travel time model of equation 5.30 is approximated because of the storage lane allocation. The two storage operations are not independent of each other due to the process of successively storing into one storage lane. Instead, there is a dependency of the second storage operation on the first, if a multiple storage is performed. This also means a violation of the Markov property. When trying to model the exact storage lane allocation with a changed Markov Chain, no solution can be found.

5.3.5 Validation

Again, we use the simulation model to validate the analytical models derived in this section. The parameter configuration for the comparison is equal to that of Table 4.7. The comparison of the analytical calculation with the results from the simulation based on the mean travel time is shown in Table 5.11.

Except from the models for multiple storage, the absolute difference is less than 0.7 seconds or around 1% in all cases.

Model	Analytical Results	Simulation Restuls	95% confidence interval		Delta abs.
			lower	upper	
NN1	62.1141	61.8796	61.8593	61.8999	-0.2345
NN2	61.7975	61.6668	61.6350	61.6987	-0.1306
NN3	61.0356	61.1702	61.1571	61.1834	0.1347
NN5	58.3466	58.6900	58.6679	58.7120	0.3434
FF	60.6394	59.9930	59.8495	59.8912	-0.6464
SL	60.2455	60.2715	60.2511	60.2919	0.0260
MS.Sim	58.9328	57.9762	58.9512	58.0012	-0.9566
MS.Suc	61.1523	58.6839	58.6656	58.7022	-2.4683

Table 5.11: Results of the validation for the analytical models from section 5.3.

Because of their approximated elements, the multiple storage models exhibit a lower validation accuracy. For both, the occurrence probability is a crucial part of the analytical model, which is composed from the probability of finding an adequate constellation of storage lanes and the generating probability of such constellations (see subsection 5.3.4). We evaluate the probability that multiple storage is possible for all filling levels with the simulation model to validate this approach. For simultaneous multiple storage, the average difference of the adjusted occurrence probability ($\hat{P}(MS.Suc)$) over all filling levels is 2.5835%, for the filling level of 90% the difference is 9.18%. The mean travel time differs by 0.6% on average across all filling levels, with a maximum deviation of 1.8 seconds (or 3%) for filling levels between 80% and 85%. These results show that the adjusted occurrence probability yields a good approximation quality for the travel time model.

For successive multiple storage, the comparison of the mean travel time with the simulation results shows a deviation of 3.57% on average for all filling levels. The highest differences (higher than 5%) are observed for filling levels below 15%, which are of minor practical relevance. This deviation arises as the storage lane allocation is changed by successive multiple storage: Especially for low filling levels, the amount of completely filled storage lanes is rather small in the original storage lane allocation. However, with multiple storage this amount is increased. As a consequence, more rearrangements are necessary which is not covered by the analytical model. For the occurrence probability ($\hat{P}(MS.Suc)$) deviations can be observed for filling levels above 70% with absolute values of up to 8%. The average deviation for all filling levels is 1.05%. For both variants, more graphical representations of these findings can be found in Figures A.5 and A.6 in the Appendix A.

5.4 Calculated Comparison

The analytical models derived in this chapter allow to easily calculate and compare the respective main routing policies. Therefore, in this subsection we provide some calculation examples with the analytical models of the main routing policies and show how these formulas can be applied. First, we present a static comparison using the parameter settings applied in section 4.5 (see Table 4.7). Subsequently, we vary different parameters to gain more insights about their influence on the different models.

5.4.1 Static results

For the static comparison, the parameter settings from Table 4.7 apply with the exception of dwell times which are set to $t_{dead} = 0.3s$ and $t_{mast} = 1s$.

Model	Probability of the execution order of the underlying model P(SSRR,SRSR)	Travel time in seconds	In % of the underlying model
NN5	$(0,1)$	69.9781	92.8%
MS.Sim	$(1,0)$	69.7597	93.5%
FF	$(0,1)$	72.2710	94.4%
SL	$(1,0)$	71.8293	96.2%
NN3	$(\frac{1}{2},\frac{1}{2})$	72.6460	96.9%
MS.Suc	$(1,0)$	72.7361	97.5%
NN2	$(\frac{1}{2},\frac{1}{2})$	73.4079	97.9%
NN1	$(\frac{1}{3},\frac{2}{3})$	73.7321	98.2%

Table 5.12: Calculated comparison of the analytical models from section 5.3.

Table 5.12 shows the computed travel time for the different analytical models and the percentage of travel time compared to the respective underlying model. We observe that NN5, simultaneous multiple storage and Flip Flop achieve the best results in terms of travel time reduction.

It is notable that successive multiple storage does not yield results comparable to the simultaneous version. The reasons are the more time consuming load handling procedure and a higher amount of dwell times for the successive version.

Overall, the potential for the reduction of travel times appears rather limited with less than 10%. However, the potential strongly depends on the applied parameter configuration, as the scaling factor $\frac{L}{v_x}$ to de-normalize the rack model is an important driver. In the present example configuration, the scaling factor is 6 (= $\frac{24m}{4m/s}$), which is relatively small. The parameter configuration applied by Lippolt (2003), as an example, yields a scaling factor of 40.83. By applying greater scaling factors, both the travel time and the potential for reduction are enhanced.

5.4.2 Comparison of all models for varied parameters

We study the behavior of the mean travel times when varying the following parameters.

1. Filling level
2. Ratio of length to height of the rack $\frac{L}{H}$, keeping $b = 1$ constant
3. Scaling factor $\frac{L}{v_x}$, keeping $b = 1$ constant

If not mentioned otherwise, again the parameter configuration from Table 4.7 applies.

Filling level

Figure 5.18 shows the behavior of the travel times depending on the filling level. The left part is calculated without dwell times ($t_{dead} = 0s$ and $t_{mast} = 0s$), while the right part is equal to the case presented above ($t_{dead} = 0.3s$ and $t_{mast} = 1s$). L/H is constant with 2, while L/v_x is constant with 6.

The behavior of the analytical models is similar in both cases. For low and medium filling levels, simultaneous multiple storage clearly outperforms all other models. Multiple storage is sensitive to high filling levels, which is why the travel time decreases stronger than those of other models. For filling levels higher than 90%, NN5 and Shortest Leg are the first to achieve lower travel times. The travel times of the remaining models are higher and lie closely together. Among the remaining models, successive multiple storage has to lowest travel time for filling levels lower than 70%, while Flip Flop has the lowest travel time for higher filling levels. Moreover, all models outperform the basic model and can improve the travel time compared to it.

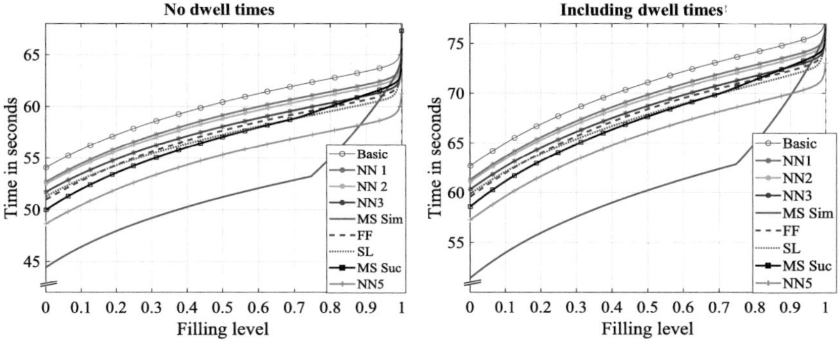

Figure 5.18: Travel times for variable values of the filling level.

Ratio of length to height

Figure 5.19 shows the behavior of the travel times for a varied ratio of the length (L) to the height (H) of the rack, L/H. The left side refers to the results with a filling level of 80%, while the right side refers to 90%. That ratio is varied between 1 and 10 by increasing L. To maintain the shape factor constant at 1, the velocity in x-direction is varied accordingly:

$$v_x = \frac{L}{6}$$

In this way, the scaling factor is constant with 6. An increase of L/H means that the rack becomes longer and has more storage positions compared to its height. However, the maximum travel time in both directions is the same.

In general, all models have a linear relation to the ratio of length to height. For a filling level of 80%, simultaneous multiple storage shows the lowest travel time in the considered range. For ratios of less than 5, NN5 ranks second, while for higher ratios successive multiple storage is in the second best position. For a filling level of 90%, NN5 and simultaneous multiple storage behave quite similar and take the lead in terms of travel time. Only for small ratios of less than 2, NN5 can achieve lower travel times.

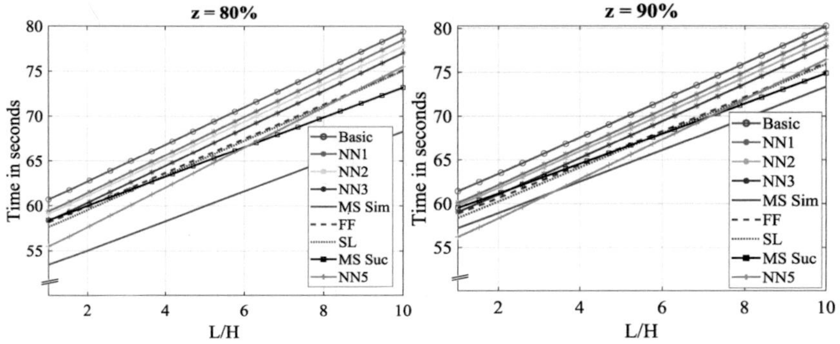

Figure 5.19: Travel times for variable values of the length to height ratio.

Scaling factor

Figure 5.20 shows the behavior of the travel times for a varied scaling factor, $\frac{L}{v_x}$. A higher scaling factor means a longer travel time to travel the maximum distance in the rack. To maintain the shape factor constant with 1, the following relation is maintained during the calculations:

$$\frac{L}{v_x} = \frac{H}{v_y}$$

The left side refers to the results for a filling level of 80%, while the right side refers to 90%.

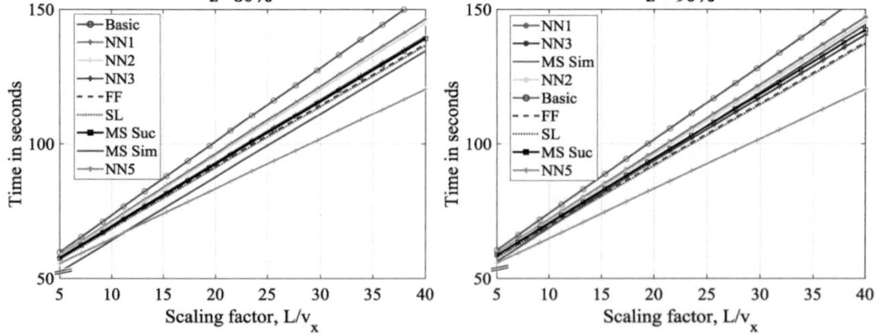

Figure 5.20: Travel times for variable values of the scaling factor.

In general, all models have a linear relation to the scaling factor. Figure 5.20 shows that the advantage of simultaneous multiple storage is limited to a range of scaling factors below 12 and the filling level of 80%. With an increase of the scaling factor, first NN5 takes the lead in travel time and subsequently increases its travel times advantage over simultaneous multiple storage and the other models. For the filling level of 90% this advantage is more distinct compared to the filling level of 80%. Compared to the gap to NN5, the travel times of the other models lie closer to each other, but increase there differences with increasing scaling factor.

When analyzing the influence of the filling level and the ratio L/H for a fixed scaling factor of 40, the findings for high scaling factors of this paragraph are confirmed. For a 80% filling level, simultaneous multiple storage ranks second while for 90% Shortest Leg ranks second. An illustration can be found in Figure A.7 in the Appendix A.

To conclude this considerations, the savings in travel time compared to the underlying model are evaluated depending on the ratio of the length to the height and the scaling factor. This is illustrated in Figure 5.21. As the scaling factor increases, the savings increase for all models. The travel time of NN5 shows the highest slope with an increase of 15% between a scaling factor of 5 and 40.

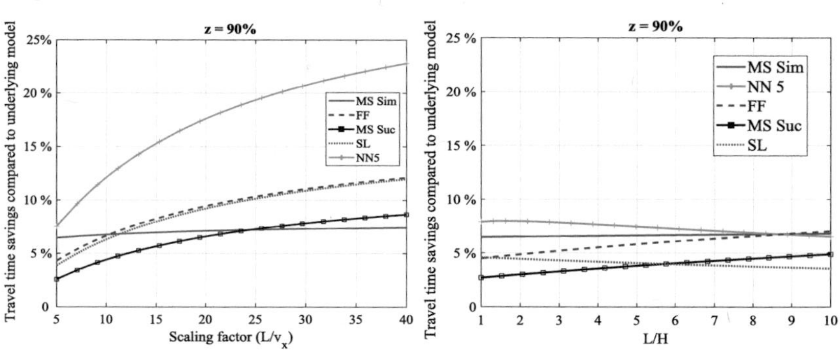

Figure 5.21: Savings compared to the underlying model depending on L/v_x and L/H.

The increase in L/H has a more diverse influence on the savings. Successive multiple storage and Flip Flop show an increase of abut 2%, while NN5 and Shortest Leg slightly decrease.

5.4.3 Computation of selected intersections

Moreover, analytical formulations allow to compute the intersections between the models. This provides more insights about the specific mutual performance for different parameters in the comparison of two models. Motivated by the observations for different scaling factors, we calculate intersections to further analyze the performance of multiple storage and NN5. Again, we use the configuration of Table 4.7 and indicate which parameters are changed for the evaluations.

First, we analyze the intersection of simultaneous multiple storage with NN5 and successive multiple storage with NN5 for the relation of filling level and scaling factor. Different scaling factors are generated by varying v_x and v_y. To maintain the shape factor constant with 1, their ratio is kept constant according to:

$$\frac{L}{H} = \frac{v_x}{v_y} = 2$$

The illustration on the left side of Figure 5.22 shows that for scaling factors higher than 15, NN5 performs better for all filling levels in general.

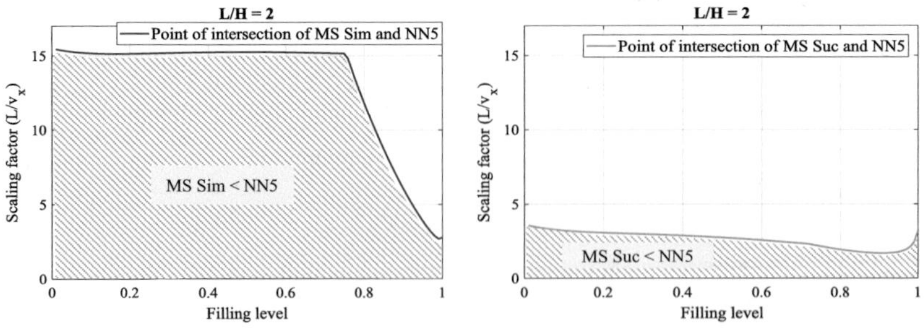

Figure 5.22: Intersection points of MS Sim and MS Suc with NN5 depending on the filling level and the scaling factor.

For filling levels up to 75% the point of intersection remains stable at this level, while it decreases significantly for higher filling levels. For high filling levels of more then 90%, simultaneous multiple storage only provides an advantage for scaling factors of less than 4. In the comparison of successive multiple storage with NN5 on the right side of Figure 5.22, the point of intersection always lies below a scaling factor of 4. Below this point, successive multiple storage performs better than NN5.

Figure 5.23 depicts the intersection of simultaneous multiple storage with NN5 and successive multiple storage with NN5 for the relation of filling level and the ratio of length to height of the rack. The length of the rack is varied and the shape factor is set to 10. To maintain the shape factor constant at 1, the velocity in x-direction is varied accordingly:

$$v_x = \frac{L}{10}$$

The left side of Figure 5.23 shows that below a filling level of about 80%, simultaneous multiple storage always has a lower travel time than NN5. With increasing filling level, the intersection occurs for higher values of L/H. Consequently, the point of intersection marks the minimum ratio at which simultaneous multiple storage still has a lower travel time than NN5, while for smaller ratios NN5 performs better. This behavior corresponds to Figure 5.21. On its right side it can be observed that the savings of NN5 decrease with increasing L/H.

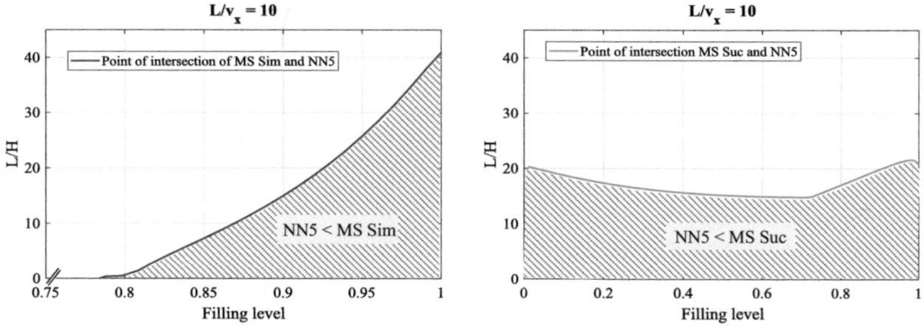

Figure 5.23: Intersection points of MS Sim and MS Suc with NN5 depending on the filling level and the length to height ratio.

On the right side of Figure 5.23 we see that the point of intersection lies in the area between a length to height ratio of 15 to 20 for all filling levels. For smaller ratios, NN5 always performs better for all filling levels. Between filling levels of 0% and 72%, the ratio of length and height at which the models intersect decreases, while for higher filling levels it increases with the filling level.

Figure 5.24 shows the intersection of simultaneous multiple storage with NN5 and successive multiple storage with NN5 for the combination of the scaling factor and L/H. For this evaluation, the filling level is set to 90%. The length of the rack is varied, while v_x and v_y are changed as well as, as long as their ratio is kept constant according to:

$$\frac{L}{H} = \frac{v_x}{v_y}$$

In this way, we maintain the shape factor constant with 1. The intersection points in Figure 5.24 indicate, for different ratios of L/H, up to which scaling factor multiple storage exhibits a better performance than NN5. If the scaling factor is higher than the value of the intercept, NN5 has a lower travel time. For both cases, a similar behavior is observed: As L/H increases, the point of intersection is reached for higher scaling factors. For values of L/H below 10, the scaling factor that marks the intersection is less than 8.

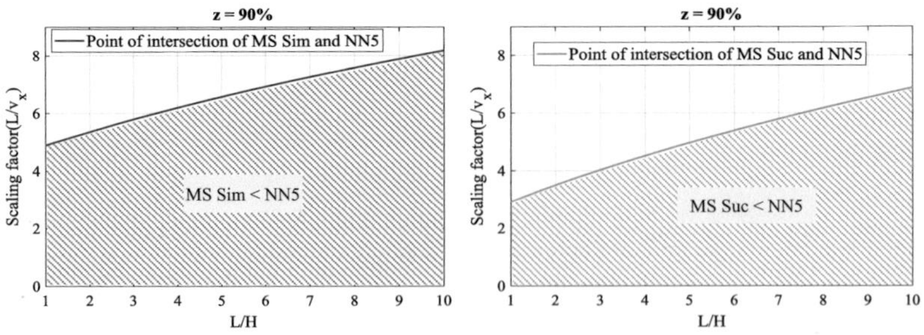

Figure 5.24: Intersection points of MS Sim and MS Suc with NN5 depending on the length to height ratio and the scaling factor.

5.4.4 Summary

This section shows some examples of performance analyses that are possible with the analytical models. However, a rack configuration consists of many different parameters and allows many different combination of parameter settings. From the variation of different parameters based on the setting of Table 4.7 we can conclude the following:

- The scaling factor that describes the maximum possible travel time within the rack is an important parameter in the comparison of the different models.
  - For small scaling factors, simultaneous multiple storage is superior compared to the other models, especially for filling levels lower than 90%.
  - For scaling factors higher than around 15, NN5 takes advantage because of the nearest neighbor principle and has the lowest travel times for all filling levels. This shows, the longer the travel times become in general the more important it is to shorten the travel between distances.
  - For filling levels of about 90%, both simultaneous and successive multiple storage can only take the advantage over NN5 for scaling factors below 8. However this is only possible for ratio of L/H of up to 10. For shorter racks, that have three or four times the length than their height, the respective scaling factor is smaller than 6.
- Generally, with increasing filling levels, the performance of multiple storage declines as the possibility of performing multiple storage operations becomes rare. NN5 is not that sensitive to an increased filling level and therefore is more beneficial for high filling levels greater than 90%.
- The analytical model for simultaneous multiple storage is a conservative approach regarding the occurrence probability. The probability is higher, if three neighboring storage lanes are considered in the computation of equation 5.20. Note that in this case, it is required

to assume that the LHD can be positioned as desired in front of the three storage lanes. However, potential uprights in the rack structure do counteract the occurrence probability. We analyze these effects in more detail in the following chapter.

- The performance of successive multiple storage is inferior compared to simultaneous multiple storage in all evaluated cases. However, the comparison to the simulation results show the largest delta for successive multiple storage, i.e., the analytical model seem to underestimate the potential.
- In most cases Flip Flop and Shortest Leg show a very similar performance and rank on third and fourth best position. Moreover, this shows that Shortest Leg is possible in the observed cases, i.e., a storage location can be found within the no-cost zone.

6 Performance Analysis using Simulation

Dance...even if you have
nowhere to do it but in
your own living room.
-B. Luhrmann

The objective of this chapter is to investigate in detail the potential for throughput improvements when applying the strategies we develop in Chapter 5 compared to a random execution of the QC. In subsection 5.4, we show that throughput improvements are possible when applying the main routing policies of the strategies and discuss the impact of the parameter choice. In this chapter, we also focus on more complex strategies, which we presume to be more powerful than the main routing policies. By being complex we mean that

- Storage or retrieval positions are not predefined. The computational effort is increased in case of free selection of storage or retrieval positions as all possible positions are assessed for the determination of every cycle.
- Several additional rules are applied in the strategy next to the main routing policy.

To investigate strategies, we use the simulation software *AnyLogic* and model one aisle of a double-deep, dual load handling AS/RS for the evaluation. The strategies are analyzed under different parameter settings (see sections 6.2.1 and 6.2.2) to get an understanding of their behavior. For the examination of the strategies, we reduce the long list from Table 5.2 to a short list of selected strategies (see section 6.1.1) in order to support evaluation and illustration of the results. Moreover, several strategies are similar and are therefore likely to generate redundant results. A diverse selection of strategies facilitates that many effects can be observed.

6.1 Implementation of the strategies in the simulation model

In this section, we explain the criteria of the strategies selection and present the resulting short list. Subsequently, we explain all adjustments made to the strategies during their implementation in the simulation model.

6.1.1 Selection of Strategies: The Short List

We formulate the following **criteria** to obtain a representative selection of strategies from the long list presented in Table 5.2.

1. Include every main routing policy.
2. Consider different degrees of complexity in terms of storage and retrieval selection (i.e. strategies that use fixed position for storage and retrieval as well as strategies with an optimized selection of positions).
3. Ensure a balanced mix of components and avoid similar strategies.
4. Balance implementation effort and value of expected insights.

Table 6.1 represents the short list of the strategies we select for the simulation studies.

We choose *three to four* strategies with different degrees of complexity from every main routing policy. In all cases, the *simple* version of the main routing policyis included. In that way, the potential within each main category of main routing policy can be evaluated.

In each category, we incorporate both strategies with and without the free selection of positions (if both are existing).

ID	Main routing policy	Additional rules and description	Resulting sequence	Storage position arbitrary	Choose retrieval requests freely	Implemented for class-based
SL1	*Shortest Leg*	No additional rules	SSRR	x	-	✓
SL2	*Shortest Leg*	+ RNN for retrievals	SSRR	x	x	✓
SL4	*Shortest Leg*	+ Using No-Cost Zone	SSRR	x	x	✓
SL5	*Shortest Leg*	+ Avoid regular rearrangements	SSRR	x	x	✓
NN1	*Nearest Neighbor*	Simple	SSRR or SRSR	-	-	✓
NN2	*Nearest Neighbor*	RNN	SSRR or SRSR	-	-	✓
NN5	*Nearest Neighbor*	Storage near retrieval	SRSR	x	-	✓
NN7	*Nearest Neighbor*	Retrieval near storage	SRSR	-	x	✓
FF1	*Flip Flop*	No additional rules	SRSR	x	-	✓
FF3	*Flip Flop*	+ RNN for retrievals	SRSR	x	-	
FF6	*Flip Flop*	+ Storage near retrieval + simple NN for R2	SRSR	x	x	
FF8	*Flip Flop*	+ Storage near retrieval + using No-Cost Zone	SRSR	x	x	
MS1	*Multiple-Storage*	No additional rules	SSRR	x	-	
MS2	*Multiple Storage*	+ Simple NN for retrievals	SSRR	x	-	
MS7	*Multiple Storage*	+ Retrieval near storage	SSRR	x	x	
MS9	*Multiple Storage*	+ Avoid regular rearrangements	SSRR	x	x	
IT1	*Increase Tango*	+ Avoid regular rearrangements	SSRR	-	x	✓
CB1	*Class based storage*	Retrieval from same class as storage	SSRR or SRSR	x	x	✓
CB2	*Class based storage*	Retrieval near storage; same class	SRSR	x	x	✓

Table 6.1: Short list of strategies that are implemented in the simulation model

Moreover, we select various *additional rules* across the groups, such as no-cost, avoiding rearrangements, different variants of nearest neighbor (simple, RNN, storage near retrieval, retrieval near storage).

Note that for the group *Increase Tango* only one strategy (IT1) is selected. From chapter 4, we know that its absolute potential for improvement is small. However we select the main routing policy to understand how forcing tango and avoiding regular rearrangements work, while the additional rules (applied in IT2 - IT4) are sufficiently present within other strategies.

Some strategies are not employed, due to their implementation effort. From the group of the *Nearest Neighbor* strategies, NN9 is not selected because pairing from all open storage locations with all retrieval requests is relatively complex, while the enhancement compared to NN5 and NN7 is presumably minor. The analysis of the nearest rearrangement position shows that the nearest available position already is the next or second-next position (on average for filling levels below 95%, see Table 4.9 and section 4.5). Pairing of open locations and retrieval requests can hardly lead to shorter distances between two positions. The same applies for the minimum perimeter rule in the *Flip Flop* group.

The category of *Class based storage* strategies represents a special case. Besides the two dedicated strategies, CB1 an CB2, we evaluate additional strategies from the selection. Depending on the way the storage positions are selected, some strategies can be transferred to class based storage assignment without any changes. Other require further control to ensure correct application. The last column of Table 6.1 shows whether the strategy is implemented for usage within class based storage assignment. NN1, NN2, NN7 and IT1 can be transferred without any modifications. For these, storage positions are predefined and thus are automatically selected from the correct class. The remaining strategies are modified to ensure the correct assignment of the storage units to the respective class (see subsection 5.2.2).

Strategies from the group *Multiple Storage* are not implemented for class based storage assignment, as the combination of both has too many limitations. The storage assignment requires that both storage units belong to the same group. As no sequencing of the physical storage units is assumed, we expect this to be very unlikely.

In total, we select 19 strategies for the examination by our simulation studies, 17 strategies are suitable for random storage assignment and 12 strategies for class based storage assignment.

6.1.2 Adjustments for Implementation of Strategies

To bring the implementation closer to practical applications, we adjust the following assumptions for the simulation model in general.

- We consider both sides of the aisle, i.e., we have two storage racks opposed to each other.
- Nearest neighbor selection is time-based, i.e., taking travel time and access time of the LHD into account.
- If both retrieval requests are randomly selected from the same storage lane and the unit at the rear is scheduled to be retrieved first, the order of the two retrieval operations is changed with the result that unit at the front is retrieved first.
- The selection of storage positions at the front of a storage lane is avoided, if an upcoming retrieval is blocked in this way.
- The storage units are no longer undefined. We differentiate between a distinct number of SKUs, which are understood as different articles or product types. They each have their own demand behavior. For each SKU, several storage units (items) exist in the storage place. Therefore, retrieval selection consists of the two aspects described on page 22 in subsection 2.2.2. When explaining the control parameters in subsection 6.2.1, we provide further details regarding retrieval selection.

In general, strategies are implemented according to their logic described in section 5.2. Adjustments are made in the following cases:

- Multiple Storage
  - Both variants of multiple storage are implemented. For successive multiple storage, '**a**' is attached to the particular strategy ID, e.g. MS1a. For simultaneous multiple storage, '**b**' is attached to the strategy ID, e.g. MS1b.
  - A pair of available storage positions is required to execute multiple storage. In the simulation, we constantly record all pairs of available storage positions in a list, and only choose storage positions from that list. As soon as the number of available pairs is less than a certain value, retrieval positions are selected in such a way that a new pair of storage positions is generated. In this way, performing a multiple storage operation is enforced in most cycles.
- Nearest Neighbor
  - NN5 and NN7 are implemented for both execution orders. For the execution order SRSR, '**a**' is attached to the strategy ID, e.g. NN5a. For the execution order SSRR, '**b**' is attached to the strategy ID, e.g. NN5b.
- All applications of the no-cost zone
  - If no appropriate position within the no-cot zone is found, the perimeter of the former no-zone is gradually enlarged until an appropriate position is found.

6.2 Set - Up of the Simulation Studies

In this section, we describe the general framework of our simulation model. Consecutively, we explain the control parameters, their settings, the performance indicators and present a general validation of the simulation method.

6.2.1 Description of Control Parameters

The simulation allows many input parameters to describe the physical and operational set-up. They are clustered in different parameter groups which we introduce in the following.

Physical Parameters

These parameters define the physical configuration of the AS/RS, i.e. they define dimensions and dynamic behavior of the system.

- Dimensions of the storage rack, i.e. height and length of the rack and the storage lanes.
- Dynamics of the S/R machine: Speed in horizontal and vertical direction, acceleration and deceleration in both directions.
- Load handling times, i.e., the time needed to access front or rear position of a storage lane.
- Dwell times such as dead times and mast damping time.
- Interval and width of uprights between storage positions.
- Location of I/O position.

Operational Parameters

There are different operational parameters that control how requests are generated, assigned to the storage place and selected for retrieval.

The parameters that specify the **allocation** of the storage rack are

- Filling level,
- Mode of storage assignment, i.e.:
  - *random* storage assignment,
  - *class based* storage assignment.

Retrieval requests are generated and recorded in a list that is described by the

- Maximum size of the list of retrieval requests.

As soon as a new retrieval request is generated, it is saved in the **list of retrieval requests**. One entry in the list represents the demand for a specific SKU, as illustrated in Figure 6.1. In the simulation model, the amount of requests in the list is regularly checked and new requests are generated until the maximum size of the list is reached.

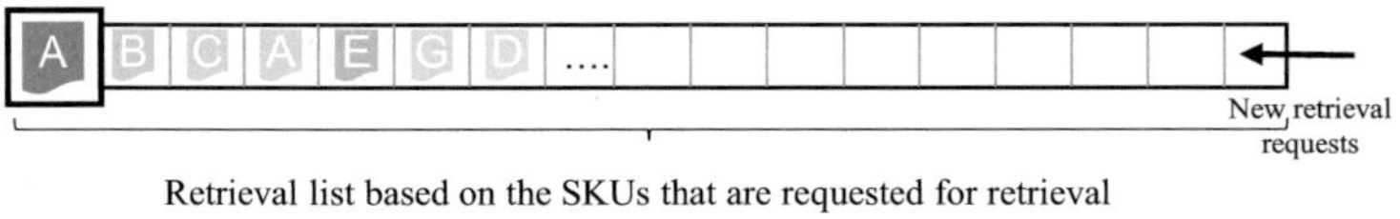

Figure 6.1: Illustration of the list where the demand for SKUs is recorded.

We consider both aspects of **retrieval selection** defined in subsection 2.2.2 and use the following parameters to describe the selection process:

- Retrieval policy for SKUs, i.e.,
  - *FCFS* selection,
  - *Free choice* from all units,
- Capacity of the sequencing window.

Regarding the first aspect of retrieval selection, we distinguish two **retrieval policies** for the selection of SKUs. Either, the longest stored unit of a given SKU type is selected for retrieval (FCFS) or there is free choice from all items of the SKU type (see Figure 6.2).

The retrieval list as well as the retrieval policies are implemented for every strategy. However, resorting of the retrieval list only occurs for strategies that require a free retrieval selection. For other strategies, the requests in the retrieval list are executed in order of their arrival into the list.

If resorting of the retrieval list applies, a control mechanism is needed to ensure that all requests are executed and delaying is prevented.

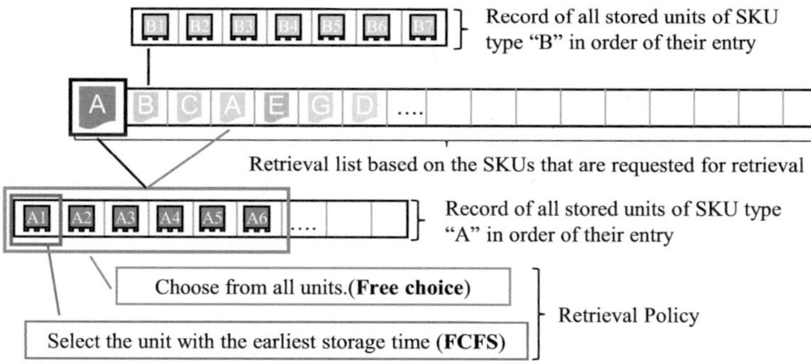

Figure 6.2: Illustration of the selection policies of SKUs.

For this purpose, we implement a **sequencing window** that represents a segment of the retrieval list and works as follows: All requests are consecutively numbered in the order of their entry into the sequencing window. The difference between the number of the first and the last request within the window is continually calculated. The maximum difference allowed is the maximum capacity of the sequencing window (this is an input value) minus one. As soon as the calculated difference equals the maximum difference allowed, no other request is permitted to enter the window. As a consequence, the remaining requests have to be executed. Figure 6.3 and Example 6.1 illustrate this behavior.

Example 6.1 *Consider a sequencing window with a capacity of 7 slots. The maximum difference allowed between the numbers of the requests is 6. For an unbroken filling from '1' to '7', the algorithm prevents number 8 from entering the list, as no free slot is available. This is shown in the upper part of Figure 6.3. In the lower part of Figure 6.3, another situation is shown: The first request within the sequencing window has the number '3', while the last request has the number '9'. The difference between the two is 6, which is the maximum difference allowed. Therefore the window must not be filled up with new requests until the request with number 3 is selected for retrieval and exits the window. As soon as the request representing number 4 takes the first slot, the request with number 10 is allowed to move up (10-4 = 6).*

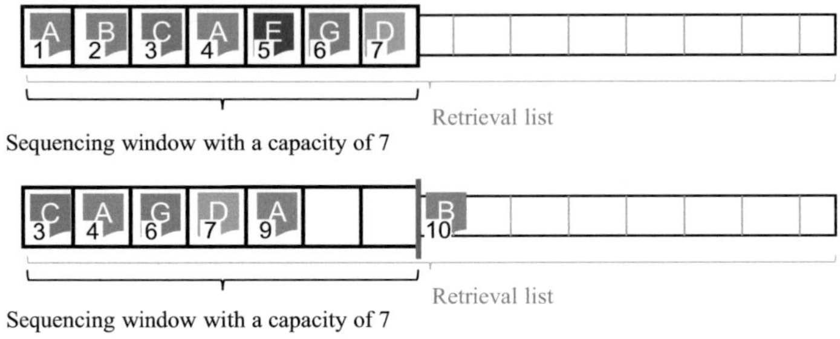

Figure 6.3: Illustration of the sequencing window

Note that without free selection of retrieval requests (no sequencing window), the retrieval policy for the SKUs has a minor relevance. Corresponding strategies use predefined retrieval positions and therefore do not have the freedom to choose from every item of a given SKU. If the retrieval policy parameter is set to *free choice,* a random item is selected.

Parameters in the context of **SKUs** are used to describe a certain SKU profile that consists of turnover frequency and demand profile. We use the following parameters.

- Number of SKU types,
- Gini coefficient of the SKUs demand distribution,
- Marginal values in order to define class limits for A and B products for class based storage .

Based on the number of SKUs and the desired Gini coefficient, the simulation determines a turnover rate for every SKU. According to this turnover rate, the simulation generates retrieval requests, i.e., SKUs with a higher turnover ratio are requested more frequently. SKUs for storage are created equally. If the storage assignment is set to class based storage, a rule for the allocation of the SKUs to a particular class is needed. The marginal values represent a specific cumulative turnover frequency to distinguish A from B, and B from C products, respectively. The classes are located according to the commonly used L-shape configuration (Roodbergen and Vis 2009, p. 350) as shown in Figure 2.7.

6.2.2 Parameter Settings

We describe different simulation experiments by setting the parameters described in the last subsection. The **physical parameters** are used to define four different configurations of AS/RSs. This means, each of these four configurations exhibits one particular parameter setting. Following real world scenarios, we choose two configurations that represent an AS/RS for pallets and two configurations that represent a miniload system . The configurations in each group differ by the rack dimensions, with one configuration for $w = 1$ and on for $w < 1$. Table 6.2 shows the settings of these four configurations in particular.

The I/O position is located at the bottom left corner in every configuration.

The operational parameters are varied within each of these four configurations. Table 6.3 shows the number of different settings and the values selected for the simulation experiments

The size of the list of retrieval requests is set to 1.5% of the occupied storage positions in all experiments. The total number of SKUs is 312 in every case. Each Gini coefficient is combined with two specific marginal values to classify the SKUs into A, B and C products. The number of SKUs in the three classes are presented in Table 6.4 for every Gini coefficient. A graphic representation of the three SKU profiles defined by the values of Gini coefficient can be found in the Appendix B (Figure B.3).

From Table 6.3 we can calculate the number of combinations that are employed for every AS/RS configuration. The number is depending on the particular strategies. For strategies that apply free selection of retrieval requests (i.e., that use a sequencing window), there are 48 (= 4x3x2x2) different combinations, otherwise there are 24.

To sum up, the parameter set-up is defined by three dimensions, which are **physical parameters** and **operational parameters** and the applied **strategy**.

[3] Varied, when applicable for the particular strategy

Physical parameter	Settings			
	Pallet scenario		Miniload scenario	
Height of the rack	30 m	30 m	12 m	12 m
Length of the rack	90 m	135 m	16 m	31.2 m
Height of one storage lane	1.5 m	1.5 m	0.4 m	0.4 m
Length of one storage lane	1.0 m	1.0 m	0.6 m	0.6 m
Speed of the AS/RS in x direction	3 m/s	3 m/s	4 m/s	4 m/s
Speed of the AS/RS in y direction	1 m/s	1 m/s	3 m/s	3 m/s
Acceleration of the AS/RS in x direction	0.4 m/s^2	0.4 m/s^2	2 m/s^2	2 m/s^2
Acceleration of the AS/RS in y direction	0.6 m/s^2	0.6 m/s^2	2 m/s^2	2 m/s^2
Shape factor of the rack	1	0.667	1	0.44
Time to access the front position	5s	5s	4.5s	4.5s
Time to access the rear position	8s	8s	5.5s	5.5s
Dead time	0.3s	0.3s	0s	0s
Mast damping time	2s	2s	1.5s	1.5s
Interval of uprights in x-direction	3	3	-	-
Width of uprights	20 cm	20 cm	-	-

Table 6.2: AS/RS configurations used for the simulation studies inspired by (Dambach Lagersysteme GmbH & Co. KG 2015) and (Gebhardt Fördertechnik Gmbh 2015)

Parameter	No. of Settings	Employed Values
Filling level	4	0.8; 0.9; 0.94; 0.98
Gini coefficient	3	0.5; 0.725; 0.825
Retrieval policy for SKUs	2	FCFS, free selection
Size of sequencing window [3]	2	6, 25

Table 6.3: Parameter settings for operational parameters that are employed for every configuration given in Table 6.2

Gini coefficient	0.5	0.725	0.825
Number of SKUs	312	312	312
Classified A	48	36	34
Classified B	81	68	56
Classified C	183	208	222

Table 6.4: SKU demand profiles used in the simulation studies

6.2.3 Performance Indicators

From each simulation run, different variables are recorded in order to analyze performance and system behavior. The **throughput** is the most important from practical point of view. It can be deducted from the mean travel time observed.

In addition, we use the **rearrangement behavior** in our investigations, which is the mean occurrence probability of regular and tango rearrangements and the mean distance to the next available rearrangement position. The composition of the travel times and the storage lane allocation are further indicators, we draw conclusions from.

6.2.4 Statistical Validation of Simulation Results

Obtaining results

A **simulation run** describes an uninterrupted simulation of 100,000 quadruple command cycles with one setting of parameters. A **simulation exper-**

iment is the ten-time replication of one simulation run with identical settings, but different seed values of the random number generator. The result of one experiment is given by the mean results from the individual simulation runs. At the beginning of each simulation run, the storage rack is randomly filled according to the storage assignment settings. We start the simulation in a filled system with an adequate pool of requests and do not crop a warm-up phase in the results as the influence is negligible (see Appendix B).

Validity of the method

To ensure the validity of the simulation results and the relevance of the implications, we address the following questions:

1. Is the number of iterations sufficient to get reliable results?
2. Is the structural influence of the random numbers generator affecting the results?

To measure the reliability of results (1.), we use the relative standard error of the sample mean which is calculated by (Kohn and Öztürk 2011, p. 71):

$$SE_x = \frac{s}{\bar{x}\sqrt{N}} \tag{6.1}$$

Here, s is the standard deviation of the sample, $\bar{x}$ the mean value of the sample and N number of observations in the sample. While s and $\bar{x}$ are evaluated in the simulation model, N is $100,000$ for every simulation run.

We use the mean of the travel time as the sample mean, as it is the most important performance indicator. In total, we conduct 53280 simulation runs with a relative standard error of less than 0.12%. Figure 6.4 shows the frequencies of the relative standard error of all simulation runs. For most runs, it lies between 0.04% and 0.06%. Therefore, we can conclude that $100,000$ are enough to obtain reliable results.

As each run is replicated 10 times with different seed values in one experiment, the deviation of the average travel times needs to be considered.

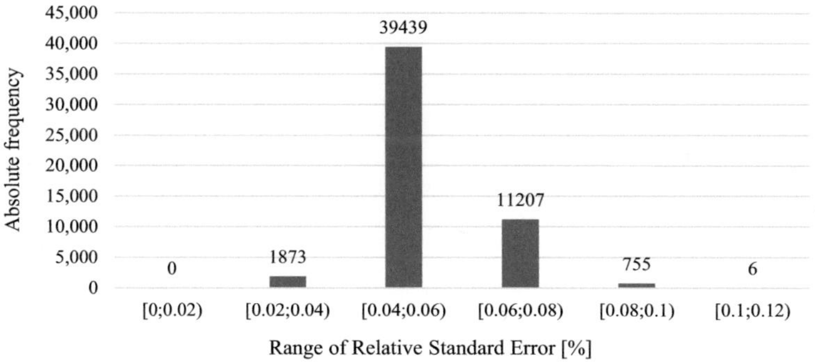

Figure 6.4: Frequencies of the relative standard error of all simulation runs in %

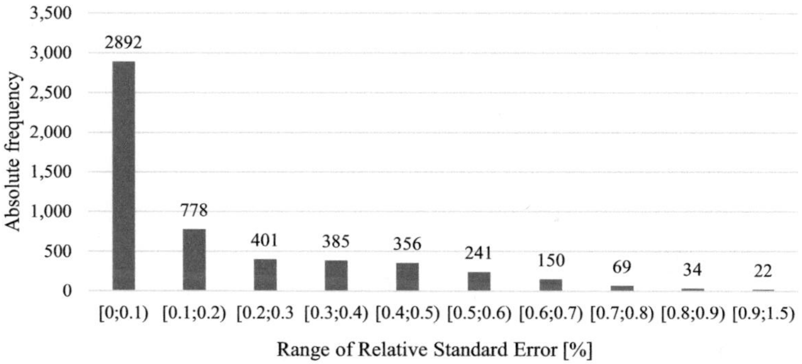

Figure 6.5: Frequencies of the relative standard error from all simulation experiments in %

To analyze the influence of the random numbers generator we calculate the relative standard error of the simulation experiments. We conducted 5328 simulation experiments consisting of 10 runs each which represent the sample size for this calculation. Figure 6.5 depicts the frequencies of the relative standard error from all experiments. Most relative standard errors are smaller than 0.1%, while the maximum observed is 1.5%. Thus,we justify that the random number generator has a minor influence.

Validity of the varied parameters

Next, we check if there is a statistically ensured influence of the parameters we apply in the simulation experiments. To ensure a significant influence, we perform an analysis of variances (ANOVA). Using the statistical software SPSS, we examine the influence of the following parameters: Strategy, Gini coefficient, size of the sequencing window, SKU selection, filling level and AS/RS configuration. We use the parameter *strategy* as a fixed factor, while the remaining parameters serve as covariates. They may skew the results and therefore serve as independent control variables. This means, as different AS/RS configurations or filling levels exhibit different initial travel time levels for the random execution (i.e., without applying a strategy), they consequently differ in travel time when applying different strategies.

Tests of Between-Subjects Effects

Dependent Variable: mean time

Storage assignment		Type III Sum of Squares	df	Mean Square	F	Sig.
random storage assignment	Corrected Model	56997115,690[a]	28	2035611.275	1828.652	0.000
	Intercept	7126.030	1	7126.030	6.402	0.011
	config	53793620.774	1	53793620.774	48324.449	0.000
	fillingLevel	90197.709	1	90197.709	81.027	0.000
	gini	75243.176	1	75243.176	67.593	0.000
	sKUSelection	333060.380	1	333060.380	299.198	0.000
	sequencingWindow	4018.702	1	4018.702	3.610	0.057
	strategy	2546343.043	23	110710.567	99.455	0.000
	Error	37904758.399	34051	1113.176		
	Total	518068225.267	34080			
	Corrected Total	94901874.089	34079			
class based storage assignment	Corrected Model	20203499,740[b]	19	1063342.091	1263.793	0.000
	Intercept	162971.397	1	162971.397	193.693	0.000
	config	19189729.471	1	19189729.471	22807.188	0.000
	filling Level	7133.177	1	7133.177	8.478	0.004
	Gini	509720.370	1	509720.370	605.808	0.000
	sKUSelection	56377.184	1	56377.184	67.005	0.000
	sequencingWindow	6065.663	1	6065.663	7.209	0.007
	strategy	420512.112	14	30036.579	35.699	0.000
	Error	16137851.278	19180	841.390		
	Total	249459802.063	19200			
	Corrected Total	36341351.016	19199			

a. R Squared = ,601 (Adjusted R Squared = ,600)

b. R Squared = ,556 (Adjusted R Squared = ,555)

Figure 6.6: Results of the analysis of variances to determine the influence of the strategies and other parameters on the mean travel time

Figure 6.6 shows the result from the ANOVA. Besides, we check the variance homogeneity (not shown) and obtain negative results. With an existing variance homogeneity, the significance limit should be set to 0.01 (Bühl 2016, p.535 f.). Except for the size of the sequencing window, we can report a significant influence of all parameters. Intuitively, this can be explained by the fact that the sequencing window can be seen as an extension of the retrieval policy (SKU selection). With a free choice from all units of a SKU, the number of potential requests to choose from increases. The same effect applies for a greater size of the sequencing window. Therefore, we do not separately analyze the size of the sequencing window in the following section.

6.3 Results: Behavior of Strategies

In this section, we present the results from our simulation experiments for all selected strategies. First, we study the results on a high level to identify which strategies achieve the best results in general. Afterwards, we further examine the influence of the varied parameters,i.e., retrieval policies, filling level and Gini coefficient and some special cases. Random execution of the cycle is indicated with the abbreviation 'x'. We always compare the pallet scenario with the miniload scenario, where, if not mentioned otherwise, the settings with the shape factor $w < 1$ is used.

6.3.1 High Level Results

Cycle time can be reduced down to 55% for random storage assignment and nearly down to 70% for class based storage assignment (within class based, i.e., compared to a randomly executed cycle in class based storage) compared to the random execution of the quadruple command cycle. These values result from observations of the pallet scenario, as in all results the observed improvement is higher for the pallet than for the miniload scenarios. The reason is a different ratio of dwell times and travel times in the two cases. Figure 6.7 shows this relation for the random execution in both scenarios. Travel times in the case of pallet storage are longer, because the speed of the S/R machine is lower as well as the distances within the rack are greater. As strategies mainly address the reduction of travel times, the pallet scenario offers a greater leverage for travel time reductions. Dwell times decrease only if a whole travel between movement is saved or the number of load handling operations is decreased.

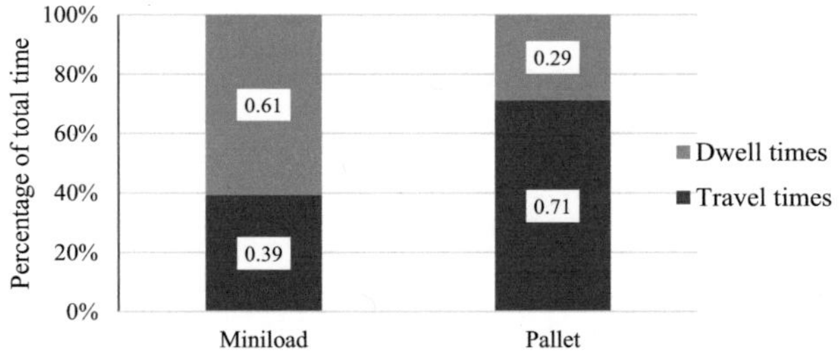

Figure 6.7: Ratio of dwell times and travel times in the two cases in % for random execution and random storage assignment

The highest improvement for pallet storage is shared between a strategy applied in random storage assignment (FF8: 69% of the randomly executed cycle) and another one for class based storage (NN7b: 85% of the randomly executed cycle within class based, 69% of the randomly executed cycle in random assignment). Both are marked with a frame in Figures 6.8 a) and 6.9 a), respectively. The reductions of travel times achieved by the leading strategies of class based storage are similar to those observed for their application in random storage assignment (FF1, NN7a, NN7b).

The highest improvement for miniload systems is achieved for random storage assignment. The lowest travel time for random storage (MS7b: 77% of the randomly executed cycle) is 9% less than the best for class based storage (NN7b: 89% of the randomly executed cycle within class based, 85% of the randomly executed cycle in random assignment). Both are marked with a frame in Figures 6.8 b) and 6.9 b), respectively. Three strategies (MS7b, FF6, FF8) perform best overall and achieve better absolute results in random storage assignment than any strategy in the class based case, but they are not implemented for class based storage. However, most strategies implemented in both scenarios can not reduce travel times below 95% of a random execution in randomized storage. In the miniload case, the potential of the strategies within class based storage is lower than in randomized storage, while in the pallet scenario the improvement potential is higher for class based storage.

Overall, strategies that are composed from several elements show a better performance, such as FF6, FF8 and MS7. These strategies reduce all travel distances in a cycle, except for one random determined movement. When exclusively comparing the main routing policies (strategy abbreviation ending with "-1"), Flip Flop performs best. The Flip Flop operation ensures that one travel between distance is saved while the others do not guarantee that. IT1 is performing worst after the randomly performed cycle, which shows that forcing tango rearrangements as the only method for improvement offers a limited potential for travel time reduction.

Next, results for all strategies in one pallet (Figure 6.8 a) and one miniload (b) scenario are presented. In both cases, the storage configuration with the greater number of storage positions is selected. Beyond that, the results from all parameters (see table 6.3) are averaged.

Results for random storage assignment

Figure 6.8 displays the mean travel times for random storage assignment. The most important observations are summarized in the following:

- FF6, FF8 and MS7b perform well in both cases.
- For pallet systems, FF8 performs best achieving a travel time of 69% of the random execution. The improvement over the second best (which is FF6) is 5%.
- For miniload systems, MS7b performs best, achieving a travel time of 77% of the random execution. The improvement over the second best (which is FF8) is 9%.
- Despite a lower potential of simultaneous multiple storage in pallet systems, the respective strategies do not fall short of their counter-

parts for successive multiple storage (see MS1b, MS2b, MS7b, MS9b in comparison to the -'a' variants in Figure 6.8 (a)). On the contrary, the results are better in all cases. This means in spite of the uprights within the rack, there are enough available adjacent positions.

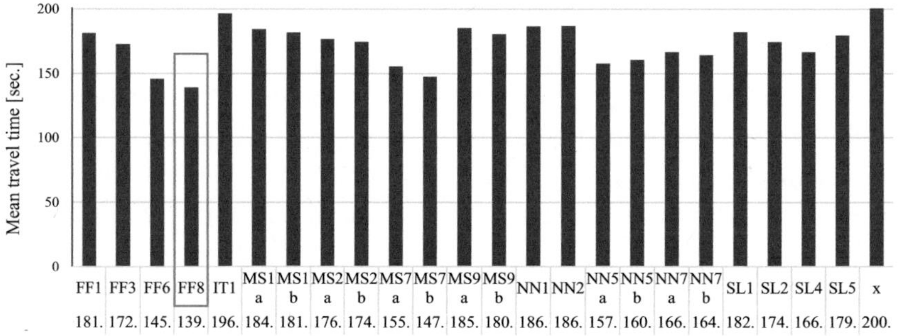

(a) Pallet case

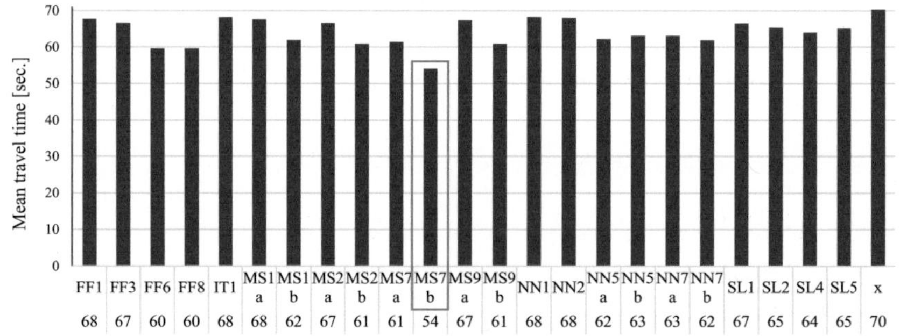

(b) Miniload case

Figure 6.8: Simulation results for Pallett vs. Miniload with random storage assignment

Results for class based storage assignment

Figure 6.9 displays the mean travel times for class-based storage assignment. The most important observations are summarized in the following:

- NN7a/b, CB2 and FF1 perform best in both cases.

- For pallet systems, NN7b achieves travel times of 85% of the random execution. For miniload systems, it achieves 89%. In both cases, the results of NN7a are slightly worse.
- Considering the dedicated class-based strategies, CB1 ranks in the lower third for both systems, while CB2 is in the upper third.

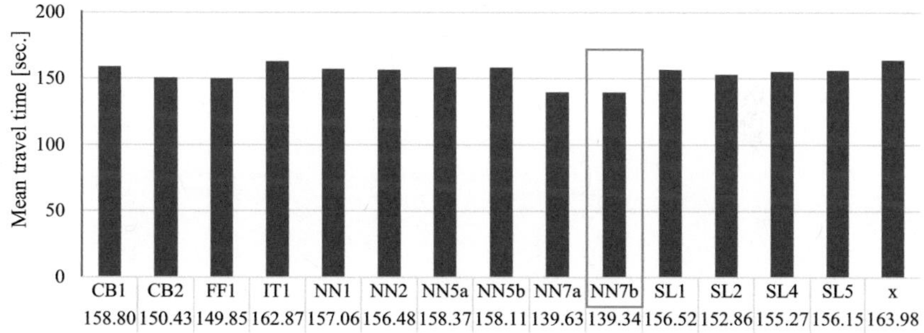

(a) Pallet case

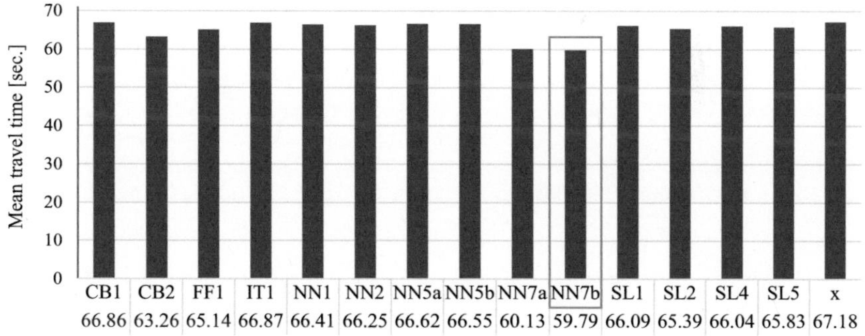

(b) Miniload case

Figure 6.9: Simulation results for Pallet vs. Miniload with class based storage assignment

Comparison of random storage and class based storage

In general, the mean travel time is decreased in class based storage assignment. In the miniload scenario, the random execution (x) with class based storage assignment attains a mean travel time of 96% compared to random storage assignment. In the pallet scenario, 82% are observed on average. However, double deep storage does moderate this effect, as random storage assignment leads to a self-acting classification. Over time, the slow-moving SKUs are moved to the rear positions of the storage lanes and because they are less requested, rearrangements decrease.

Figure 6.10 shows the proportion of SKU's of type A and C at front and rear positions of a storage lane. The left part of the diagram illustrates random storage assignment, while the right part shows class based storage.

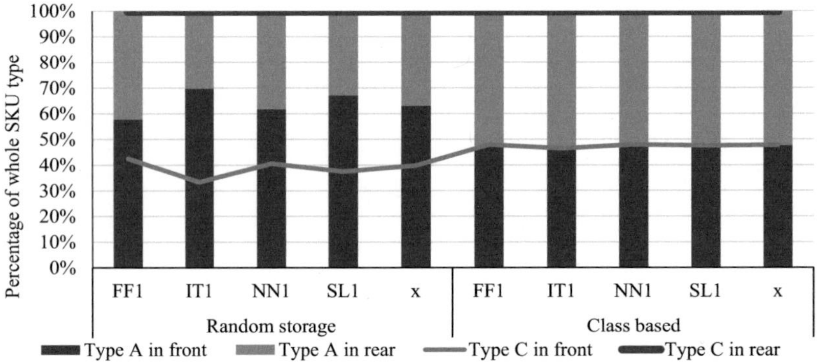

Figure 6.10: Comparison of the distribution among front and rear positions for SKU types A and C between random and class based storage assignment in the pallet scenario

When random storage is applied, type A SKU's occupy a higher proportion of front positions, while Type C SKU's are more likely to occupy rear positions. In class based storage, when these types are assigned to dedicated areas within the rack, each type takes the same ratio of front and rear positions.

This fact has direct impact on rearrangements, as shown in Figure 6.11. For all presented strategies, the ratio of rearrangements is higher in the class based scenario.This means, part of the advantage of class based storage is counteracted by a **naturally induced classification** for random storage assignment. Note that with increasing imbalance of fast and slow moving SKU's (increasing Gini coefficient), the effect increases as well.

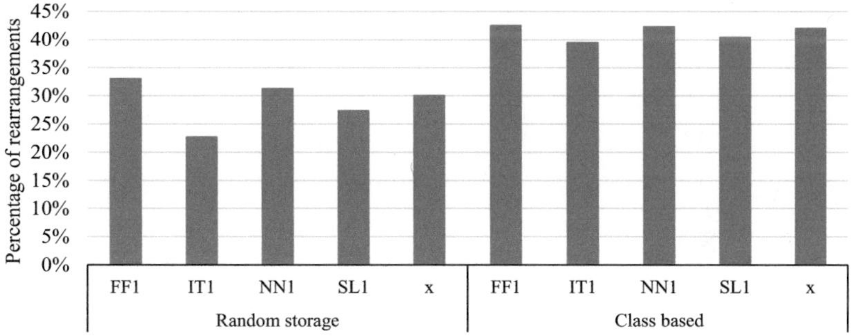

Figure 6.11: Comparison of the proportion of rearrangements (in % of all retrieval operations) between random and class based storage in the pallet scenario.

Summary of overall results

Table 6.5 summarizes the results for the top three strategies for both random and class based storage assignment. We present the observed travel time expressed as percentage of the travel time for random execution in the particular storage assignment. For class based storage assignment, the comparison is drawn to the random execution within the class based scenario. Moreover, the comparison to the random execution in random storage assignment is shown by the parenthetic numbers. All values are obtained by averaging the results from the reaming parameters from Table 6.3, i.e., filling level, Gini coefficient, retrieval policy and size of the sequencing window. They serve as benchmark for the following investigations. We assess the general validity of this observation by analyzing their sensitivity to changed parameters in the following subsections.

Storage assignment	Top 3 Strategies	% of the travel time for random execution	
		Pallet	Miniload
Random	MS7b	74%	77%
	FF8	69%	85%
	FF6	73%	85%
Class based	NN7b	85% (69%)	89% (85%)
	NN7a	85% (70%)	89% (85%)
	CB2	92% (75%)	94% (90%)

Table 6.5: Top three strategies for random storage and class based storage assignment averaged for all parameters. The parenthetic numbers show the comparison to random storage assignment.

6.3.2 Influence of the options for selection of retrieval requests

In this subsection, we study the influence of retrieval selection in order to understand how a *free choice* from all units of a given SKU influences the results compared to the *FCFS* selection of units. In general, the mean travel time is increased for the FCFS selection compared to a free choice. We can identify **two reasons** for this observation:

1. There is an increased amount of rearrangements for FCFS selection.
2. There are more possibilities to choose the retrieval unit with free selection.

The first reason applies to all strategies, while the second only affects strategies that use a sequencing window. We can detect the difference in Figure 6.12 that contrasts the mean travel times for free choice and FCFS selection. While FF6, FF8, MS7a/b, NN7a/b and SL4 show a difference of up to 45% (FF8), the reaming strategies change by 1% to 3%. Results of the miniload system show similar effects and can be found in the Appendix B (see Figures B.4, B.5 and B.6).

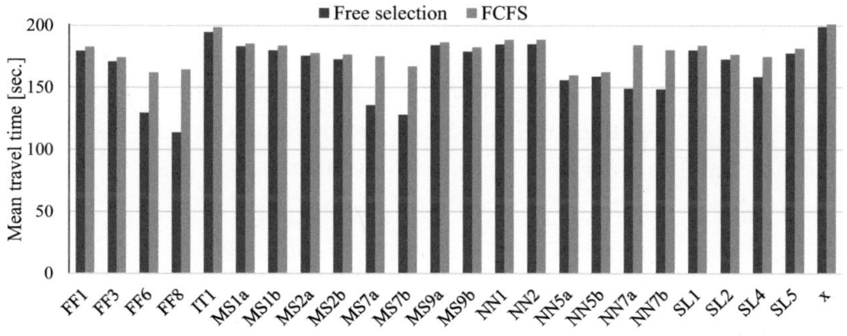

Figure 6.12: Comparison of simulation results for different retrieval options in the pallet scenario with random storage assignment

The increased amount of rearrangement, i.e., the first reason, can be explained as follows: At *FCFS* selection, the item with the longest duration of storing is selected for retrieval. At the same time, units that have a longer duration of storage are more likely to be located at a rear position of a storage lane. In general, units stored at the front position of a storage lane, are more likely to be involved in rearrangements, because they block another unit. As soon as an unit approaches a position at the rear of a storage lane (e.g. after being rearranged), it keeps that position until it is retrieved. When a unit at the rear is finally retrieved, it is are more likely to cause rearrangements (Remember the impact on slow moving SKU's in random storage assignment shown in Figure 6.10). Figure 6.13 shows that we can observe this effect in every strategy.

To explain the second reason for an increased travel time under *FCFS* selection, consider the following: With a free choice of units, strategies requiring a free selection of retrieval requests and thus a sequencing window, have **more possibilities** to find a unit that corresponds to the applied logic. For example, in case of the shortest leg policy, the a greater choice of units increases the chance to find a unit stored within the no-cost zone. The same applies for strategies that avoid regular rearrangement, as it becomes likely to find a unit at the front position. Additionally, a connection to the first reason is observed for strategies that avoid regular rearrangement: A rearrangement is forced with the first retrieval, while the selection of a unit from a front position is desired for the second retrieval. Selecting a unit from a front position can not always be accomplished with FCFS selection, because, on the one hand oldest units are rather located at the rear and on the other hand, there is a smaller range of units to select from.

From Figure 6.13, we can deduce another relation of the second reason and the first reason: Strategies with **retrieval near storage** selection show the strongest increase of the rearrangement amount (e.g. an increase of 191% for FF6 in the pallet scenario or 307% for NN7b in the miniload scenario). The nearest neighbor logic chooses a retrieval unit with the shortest total travel time. As load handling times are included in that calculation, units at front positions are preferred over units at rear positions. With a great number of units to choose from, most likely a unit at the front is selected. Consequently, less rearrangements occur with a *free choice* of all units from a given SKU. On the contrast, with *FCFS* selection, nearest (longest stored) units are more likely to be at the rear of a storage lane. For all strategies with **retrieval near storage** selection (FF6, FF8, MS7a/b, NN7a/b), we therefore observe a strong increase in rearrangements. We conclude, FCFS selection restricts the power of these strategies.

Strategies without a sequencing window, that randomly select retrieval units, remain unaffected by the second reason.

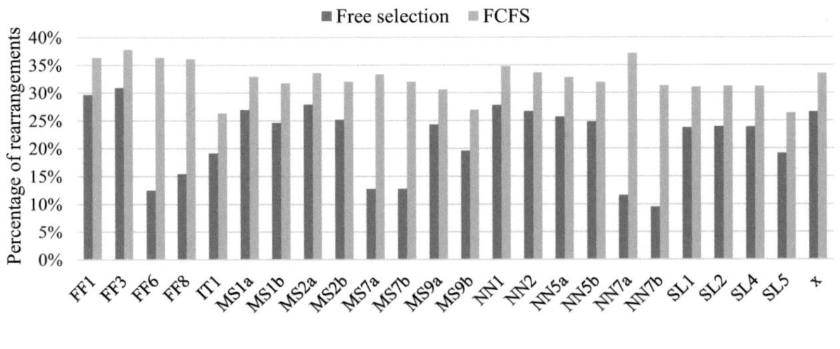

Figure 6.13: Comparison of the amount of rearrangements for different retrieval options - Pallet scenario with random storage assignment

Summary

Taking the sequencing options into account, we compare the results against those in Table 6.5 and conclude the following:

- Random storage assignment (see Figures 6.12 and Appendix B.4)
  - For *free choice* of units, we still find MS7b, FF6 and FF8 to perform well in the pallet scenario. A greater range of retrieval options increases the throughput potential so that FF8 improves down to 57% of the travel time observed for random execution.
  - For FCFS selection of SKUs, FF6 maintains its position but is outperformed by NN5a and NN5b. They show an decrease in travel time down to 79% of the travel time observed for random execution in the pallet scenario. With more restrictions regarding the retrieval unit, strategies such as NN5a/b that achieve improvements via selection of storage positions show stronger potential to decrease the mean travel time.
  - For miniload systems, MS2b and MS9b follow on MS7b with an almost equal potential of 87%. With no uprights in miniload systems, simultaneous multiple storing performs better compared to pallet systems.
- Class based storage assignment (compare Appendix Figures B.5 and B.6)
  - While the ranking is nearly unchanged compared to Table 6.5, we observe interesting results of NN7a/b that perform well at *free choice* and *FCFS selection.* Although NN7a/b decline most between the two cases, i.e., more than 20% from free choice to FCFS selection, they keep the top position with FCFS selection.
  - CB2 drops one place in FCFS selection compared to free choice of units. Instead, FF1 ranks third at FCFS selection.
  - Overall, no more than a reduction down to 91% of the travel time of the random executed cycle within class based storage assignment is achieved for FCFS selection.

6.3.3 Influence of the filling level

In this subsection, we study the change in travel time when varying the filling level as presented in Table 6.3, i.e., from 80% to 98%. Figure 6.14 shows the mean travel times for all strategies broken down to the four different filling levels in the random storage, pallet scenario. Figures showing the results for miniload systems as well as class based storage can be found in Appendix B (Figures B.7, B.8 and B.9).

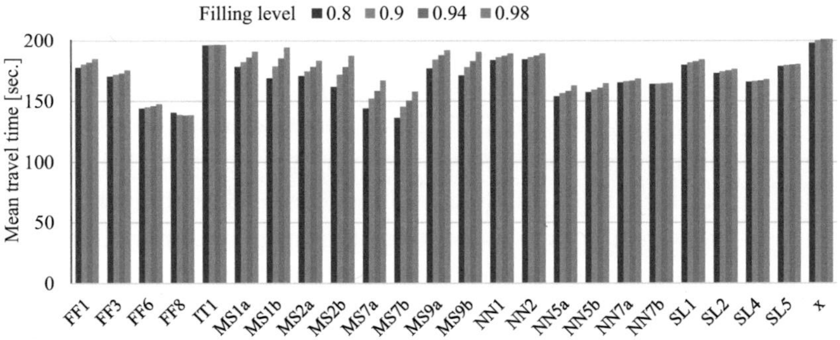

Figure 6.14: Comparison of simulation results for different filling levels in the pallet scenario with random storage assignment

The random execution (**x**) shows a travel time increase of 3% or 4%, when increasing the filling level from 80% to 98%. This observation is in line with the anticipated effect caused by rearrangements. The more units are stored in the storage place, the more likely rearrangements become. Additionally the mean rearrangement distance grows with an increasing filling level. A different behavior is observed in the following cases:

Strategies applying **simultaneous multiple storage (MS...b) react most sensitive** to an increase of the filling level. With a higher filling level, there are fewer possibilities to perform the simultaneous multiple storage operation. On average, the travel time increases by 16% for 'MS...b' strategies, when increasing the filling level from 80% to 98%. Despite uprights in the pallet scenario rack, the effect is almost identical in miniload and pallet systems. In contrast, successive multiple storage strategies do not decline in that way. Figure 6.15 shows the *number of cycles* in which no appropriate positions for multiple storage are available and underlines these findings. When comparing identical filling levels, the *number* is on a higher level for simultaneous multiple storage (i.e., MS...b; purple shaded in Figure 6.15) than for successive multiple storage (MS...a). Moreover, the *number* of cycles in which simultaneous multiple storage is not possible is higher for pallet systems than for miniload systems, but the increase in absolute numbers is very similar. For successive multiple storage, pallet and miniload scenario show the same behavior.

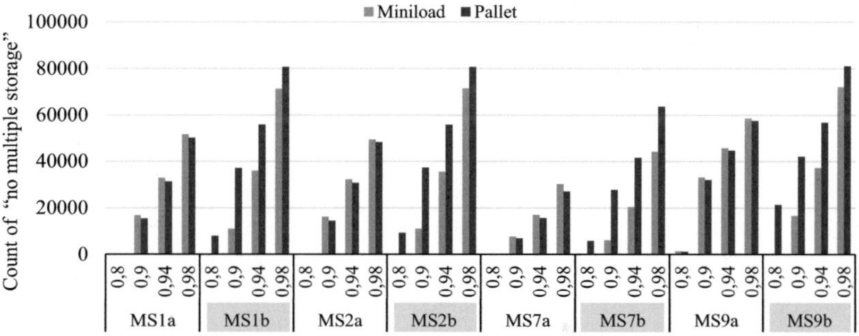

Figure 6.15: Number of cycles in which multiple storage is not possible for the filling levels 0.8, 0.9, 0.84 and 0.98

IT1 remains stable in travel time when increasing the filling level, because IT1 directly influences the occurrence of rearrangements. For the first retrieval within a cycle, a rearrangement, i.e., tango in this case, is forced and conducted whenever a blocked retrieval request is found. In the same way, the second retrieval is chosen without causing a rearrangement. By doing so, the travel time is stabilized across the filling levels. Figure 6.16 shows that IT1 causes only a slight increase in the occurrence of rearrangements. Regular rearrangements are almost completely avoided. The mean number of regular rearrangements is 286 for the filling level 98%, which is less than 1% for 100,000 cycles.

FF8 shows a decrease of 1% in travel time, when increasing the filling level from 80% to 98%. This is explained by the combination of its additional rules. The first retrieval unit is chosen near to the first storage position (*retrieval near storage*), while the second retrieval unit is chosen from the no-cost zone during the return path to the I/O position. With a higher filling level, there are more occupied storage positions. For both additional rules, there is a higher probability of finding an appropriate unit for a higher filling level which has a decreasing effect on the travel time. On the contrary, Figure 6.16 shows that for FF8 the number of rearrangements as well as the rearrangement distance increase with an increase of the filling level. This counteracts the previously explained decreasing effect. The results from **SL4** show, that a sole retrieval selection within the no-cost zone does not achieve a travel time decrease, as observed for FF8. However, the travel time in SL4 increases by only 1% with an increasing filling level.

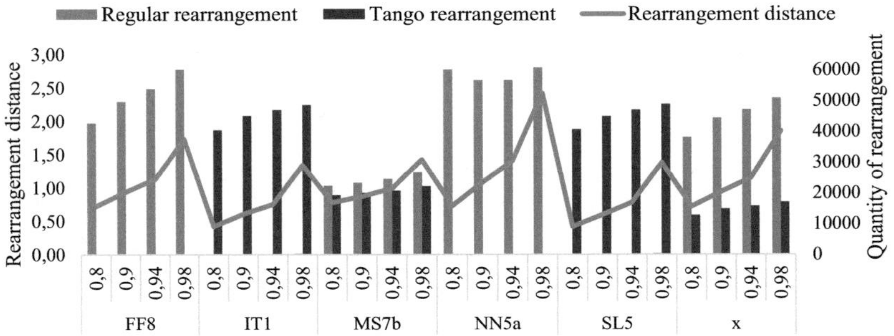

Figure 6.16: Simulation results of the rearrangement behavior for different filling level in the pallet scenario with random storage assignment

We can observe from Figure 6.16 that other strategies with the intention to avoid regular rearrangements, e.g. SL5, show a similar behavior than IT1.

Moreover, **NN5a has the lowest increase in the amount of rearrangements** and therefore is more resistant to an increase of the filling level (also applies to NN5b, but is not shown in Figure 6.16). The strategies based on *storage near retrieval* produce storage lanes in state F, which leads to a relatively low number of half-filled storage lanes. For a filling level of 80%, a rather high number of rearrangements is needed for this reason. However, the number of rearrangements remains stable with an increasing filling level. The rearrangement distance increases strongly compared to other strategies, as, in case of a rearrangement, the nearest available position has already been used for storing.

Summary

When comparing the results for averaged (Table 6.5) and varying filling levels, we make the following observation:

- Random storage assignment (compare Figures 6.14 and Appendix B.7):
  - The top three strategies are identical in the pallet scenario.
  - Although the travel time of MS7b increase sharpest from a filling level of 80% to 98%, MS7b consistently remains among the three well performing strategies with the respect to mean travel times. At a filling level of 80%, MS7b achieves 69% of the travel time for random execution in the pallet scenario and ranks first. FF8 can reach that result for filling levels of 90% or higher and takes the lead position in that range.
  - For low filling levels (80% and 90%), the advantage of multiple storage appears in the miniload scenario. MS2b and MS9b provide slightly better results (i.e., around 3%) than FF6 and FF8. For higher filling levels, we observe the same results as shown in Table 6.5. MS7b still holds the first position in the miniload case, but its advantage decreases with an increase of the filling level.
- Class based storage assignment (compare Figures B.8 and B.9 in the Appendix):
  - In the case of class based storage assignment, the results remain unchanged compared to Table 6.5.
  - All of the leading strategies show a nearly unchanged improvement compared to the random execution. This means, for all

filling levels discussed, they obtain the values presented in Table 6.5.

To conclude this subsection, the following points are the major takeaways:

- Multiple storage is unexpectedly robust towards an increase of the filling level.
- FF8 shows that it is possible to decrease travel time with an increase of the filling level, even though the number rearrangements increases.
- Strategies, that force the selection of retrieval units in order to (not) cause rearrangements can maintain their performance at all filling levels.

6.3.4 Influence of the SKUs' Gini coefficicent

In this subsection, we study the effect of different Gini coefficients of the SKU turnover distribution on the mean travel time. Figure 6.17 shows the travel times obtained in the pallet scenario with random storage assignment. Figure B.10 in the Appendix B presents the results for the miniload case.

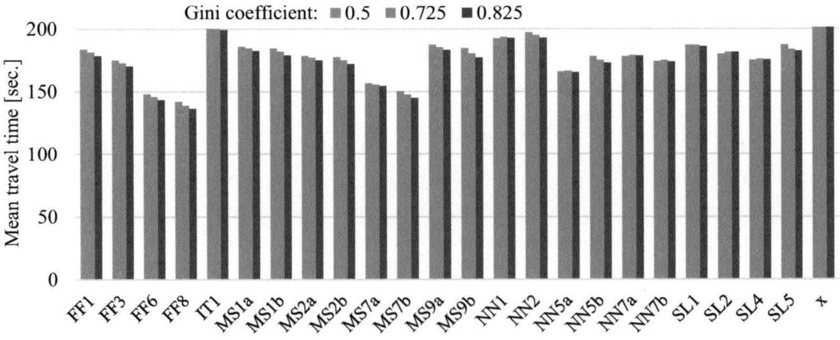

Figure 6.17: Comparison of simulation results for different Gini coefficients of the SKU distribution in the pallet scenario with random storage assignment

In random storage assignment, the impact of different Gini coefficients on the mean travel time is small. In general, travel times decrease between 0% and 5% with increasing Gini. The travel time of the random execution maintains the same level for all Gini coefficients, which means the naturally induced classification does not intensify significantly (see subsection 6.3.1) when increasing the Gini coefficient from 50% to 82.5%. However, the **general decrease of rearrangements** with an increasing Gini coefficient for all strategies indicates that a shift of slow moving SKU's towards rear positions is taking place. Strategies that show the greatest decline in travel time, also show a greater decline of rearrangements ('FF..' and 'MS..b' strategies). We identify two different reasons: First, those strategies exhibit a poor rearrangement behavior for the 50% Gini case and, by increasing the Gini coefficient, they approach an average occurrence of rearrangements with the highest Gini. Second, strategies with **simultaneous multiple storage** (MS..b) raise the number of half-filled storage lanes with an increasing Gini coefficient. A higher number of half-filled storage lanes counteracts the number of rearrangements and therefore has a declining effect on the travel time.

In class based storage assignment, the effect of SKU classification, when varying the Gini coefficient, becomes visible: With an increasing Gini coefficient, we find a **higher demand concentration** on a smaller number of SKU's. This has a positive effect on travel times, because travels of the S/R machine become more focused on the A class. Figure 6.18 shows the results for the pallet scenario in the class based storage assignment, the results for the miniload system are presented in the Appendix B, Figure B.11. Both in the miniload and the pallet scenario, the change in travel time for varying Gini coefficients is **very stable across all strategies**. For the miniload system, the travel time decreases by 6% and for the pallet system, it decreases by 17% on average, when increasing the Gini coefficient from 50% to 82.5%. If the travel distance is shortened, but the procedure of the strategies itself does not change (i.e., number of load handlings and number of travels), the relation of travel times and dwell times changes accordingly. We observe a constant decrease of the travel time ratio across all strategies for both scenarios (miniload and pallet). At the same time, the ratio of dwell times increases.

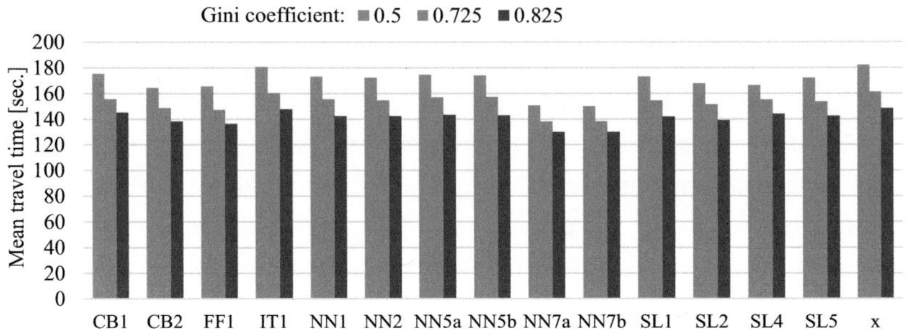

Figure 6.18: Comparison of simulation results for different Gini coefficients of the SKU distribution in the pallet scenario with class based storage assignment

Summary

We consider the results from Table 6.5 in comparison to a break-down into different Gini coefficients:

- For random storage assignment, we do not observe any differences in the placement of travel times achieved, thus results are equal to those of Table 6.5.
- Similarly, with class based storage assignment we do not report any changes among top performers.

We can summarize that the influence of different Gini coefficients (i.e., different demand distributions of the SKU's) on the strategies is mainly limited to class based storage assignment. The change in mean travel times is mainly tied to a greater concentration on high turnover SKU's and therefore observed in class based storage assignment, only.

6.3.5 Selected Cases

Strategies without free selection of retrieval requests

We distinguish between strategies that require a free selection of retrieval requests, i.e., using a sequencing window, and those that do not. In the results from the past sections, strategies using sequencing a window constantly perform best. However, we want to analyze which strategies perform well, if no sequencing of requests is possible (e.g. for technical reasons of the AS/RS). Table 6.6 summarizes the results for this case.

Storage assignment	Top 3 Strategies	% of the travel time for random execution	
		Pallet	Miniload
Random	NN5a	79%	88%
	NN5b	80%	90%
	FF3	86%	95%
	MS2b	87%	87%
	MS1b	91%	88%
Class based	FF1	91% (75%)	97% (93%)
	SL2	93% (76%)	97% (93%)

Table 6.6: Top three strategies for random storage and class based storage assignment averaged for all parameters. The parenthetic numbers show the comparison to random execution in radnom storage assignment.

For the pallet scenario with random storage assignment, the best result is observed for NN5a with a mean travel time of 78% compared to random execution. NN5b and FF3 are the second and third best strategies. In the miniload scenario, the multiple storage strategies are superior. MS1b, MS2b and NN5a achieve almost identical mean travel times between 87% and 88% compared to random execution.

In class based storage assignment, reductions of the mean travel time compared random execution are low on average. Both for pallet and miniload systems, FF1 performs best with 91% and 97% of the travel time of the random execution, respectively.

Rearrangement behavior

In a final step, we study the rearrangement behavior in general. Figure 6.19 illustrates the ratio of regular and tango rearrangements by their total quantities and the mean rearrangement distance.

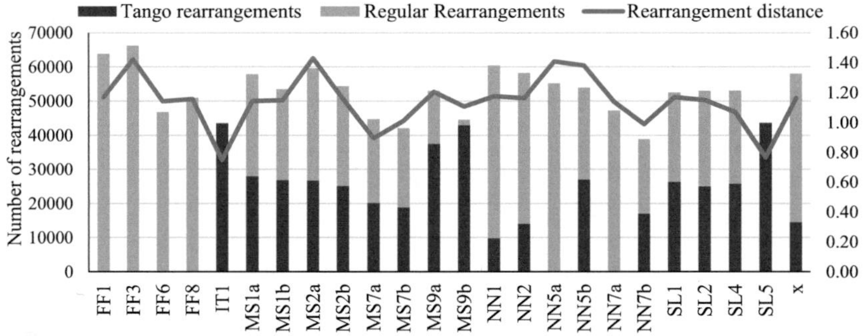

Figure 6.19: Rearrangement behavior averaged across all storage systems and all parameters in random storage assignment.

The ratio of regular and tango rearrangements occurs as expected. Strategies operating in the SRSR execution order do not perform tango rearrangements, while strategies operating in the SSRR order perform an approximate equal amount of regular and tango rearrangements. Strategies that **avoid regular rearrangements** perform almost only tango (IT1, MS9, SL5).

The top performing strategies in random storage assignment (compare Table 6.5) do not obtain the best results in terms of rearrangement behavior, but their results lie in the upper half.

Strategies that have both a small number of rearrangements and a short rearrangement distance are IT1, MS7a MS7b, NN7b and SL5. IT1 and SL5 exhibit a stable rearrangement behavior because of the avoidance of regular rearrangements. For the others (MS7a MS7b and NN7b), two successive

retrieval near retrieval operations are the reason: The second retrieval is selected as the nearest retrieval request from the first retrieval. In case a rearrangement is taking place for the second retrieval, the (formerly) storage location of the first retrieval request provides a nearby rearrangement position. The other way round, NN5a and NN5b show a high rearrangement distance due to **storage near retrieval**. In this case, the nearest rearrangement position is already occupied and can not be used (see also subsection 6.3.3).

6.4 Conclusion: Performance of Operating Strategies for QC

6.4.1 Implications of the Simulation Results

Various strategies for the execution of a QC have been examined by simulation studies. We show that the use of strategies significantly improves throughput. From the presented results we derive a number of implications:

- If no FCFS selection of items (within a SKU type) is required, a free selection should always be preferred in order to optimize throughput.
- High filling levels do not necessarily lead to increased travel times. On the contrary, some strategies can make use of a highly filled rack.
- Strategies that employ multiple storage are worth considering, even in a rack structure with uprights. Simultaneous multiple storage consistently outperforms the successive version although it faces more physical restrictions. If the execution of simultaneous multiple storage can be enabled more often, even higher potentials for travel time reduction result (on average multiple storage was not possible in 11% and 36% of the cycles at a filing level of 90% in the miniload and pallet case, respectively).
- While the most simple strategies with fixed storage and retrieval positions (NN1 and NN2) show rather smaller potential (less than 10%),

they decrease the mean travel time without substantial programming effort. They can offer greater potential in batch processing, when a greater number of cycles is optimized at once, or for higher command cycles (such as sextuple or octuple command cycles).

- The policy ***storage near...*** offers a good trade-off between implementation effort and improvement potential (e.g. NN5). Without the demand for sequencing of retrieval requests, a considerable decrease in travel time can be achieved. A combination with class based storage assignment is possible, if restriction for the selection of storage positions are kept in mind.
- Forcing the execution of tango (IT1) does not provide a major advantage in the considered cases. However, the idea to specificity select blocked and non-blocked requests works and, for example, is a valid method to control rearrangements and thus stabilize travel time, especially for high filling levels.

The simulation studies revealed many insights which serve as starting point for further investigations. Therefore, we suggest to consider these topics in more details:

- Due to double deep storage, another class based storage assignment should be considered. Instead of using L-shaped zones, slow moving SKUs should be assigned to the rear positions of storage lanes (C - classed, B - classed when required), whereas SKUs from class A can be assigned to the front positions. This method corresponds to the shift of C items to the rear positions, as observed the in case of random storage.
- In our simulation results, the best-performing strategies stand out as they effectively reduce travel distances, but not because of their rearrangement behavior. Targeted influence on the rearrangement behavior therefore may provide potential for further improvement.

6.4.2 Comparison between Analytical and Simulation Results

The findings from the evaluation of the analytical models presented in section 5.4 correspond to the results from the simulation studies. Consistently, both evaluations show that simultaneous multiple storage and NN5 are efficient strategies. Especially for lower filling levels simultaneous multiple storage performs well while NN5 can keep a stable performance for higher filling levels. For larger scaling factors, which is represented by the pallet scenario in this chapter, NN5 shows the lowest travel time, even with a filing level of 80%.

Moreover, consistent with the results from section 5.4 is that FF1 and SL1 follow at third and fourth position when considering the main routing policies only.

The simulation studies show that beyond the main routing policies, more complex strategies that are constituted from different elements can outperform simultaneous multiple storage and NN5. The main routing policies Flip Flop and Shortest Leg provide many opportunities to develop efficient strategies, such as FF8 or SL5.

7 Conclusion

Life is like riding a bicycle.
To keep your balance,
you must keep moving.
-A. Einstein

AS/RSs with dual load handling capacity that operate in double deep storage environments are frequently used in practice, although no universal approach for analytical calculations has been established yet. In this chapter we summarize the major achievements of our research.

In a first step, we review the results draw a conclusion regarding the implications for application in practice. In a second step, we give a short outlook on open topics and next steps we suggest for further research.

7.1 Summary of the Thesis

This thesis is the first to present an analytical travel time model for double-deep, dual load handling AS/RSs that takes all relevant characteristics into account and provides both definition and evaluation of operating strategies. In terms of the formulated research segment, the following results are achieved:

First segment We have presented an analytical model for travel time determination of a quadruple command cycle in double deep storage environments that does not contain crucial simplifications. It is based on common assumptions following established models, i.e. randomized storage assignment and random selection of storage and retrieval positions. Moreover, the possibilities of the considered systems are incorporated, especially with respect to rearrangements. If possible, rearrangements are performed by means of the load handling devices, which we have defined as tango rearrangement. Regular rearrangements, i.e. rearranging into an additional storage lane, are also taken into account. The storage lane allocation is modeled as a stochastic process. The stationary allocation of the process allows a determination of the rearrangement probability. The filling level and the probability distribution of the execution order are required as input parameters to obtain accurate results. The approach also permits a totally random execution of the cycle, which ensures, on the one hand, to be aligned to other general models and, on the other hand, to represent a baseline model for comparisons with advanced strategies of execution. Additionally, the model can be easily transferred to other technical implementations of dual load handling, such as the positioning above each other. Using a simulation model, we have validated our results and have been able to present deviations of the total travel time clearly below 1%. Compared to dual command cycles in double deep storage environments, throughput can be increased by at least 20% with the quadruple command cycle. By comparisons of selected execution orders, we have shown that the use of tango rearrangements mostly leads to lower total travel times. Possible obstacles are long access times of the LHD to reach the rear position. Benefits arise from a higher ratio of half-filled storage lanes and a stable rearrangement behavior, especially for very high filling levels. The shortened variant of the tango allows to further improve travel times.

Second segment Next, we have developed a list of more than 40 different routing and sequencing strategies that can be used to execute a quadruple command cycle. We have gathered existing policies from literature that are often formulated for traditional AS/RSs, have identified *main routing policies* representing different rules for operation and have adjusted them for application in quadruple command cycles where necessary. The different concepts have been combined by applying them to distinct parts within a quadruple command cycle to establish a long list of strategies. Exact and approximated travel time models for *Nearest Neighbor, Flip Flop, Shortest Leg* and *Multiple Storage* have been derived based on the general model from the first research segment and verified by means of simulation. In a last step, we have analyzed the influence of different parameters, such as filling level or scaling factor, on the travel time of the main routing policies. Among these, nearest neighbor selection of storage positions and simultaneous multiple storage exhibit the lowest travel time in general.

To further evaluate the performance, we have implemented a selection of 19 strategies in our simulation model. We have considered a simulation set-up with different AS/RS configurations for pallet and miniload cases, different filling levels, sequencing options and demand structures of SKUs. We have shown that travel time reductions of more than 40% compared to a random execution are possible. Especially combining two or more main routing policies and when applying strategies that allow sequencing of retrieval requests, considerable savings in travel times can be achieved. We have found the Flip Flop heuristic combined with both storage selection near the Flip Flop position and retrieval selection in the no-cost zone, as well as multiple storage combined with nearest neighbor selection of retrieval requests to be the most efficient ones. Within a scenario of class based storage assignment, the strategies show less potential for travel time reduction overall. Nevertheless, the nearest neighbor selection of retrieval requests has shown notable potential for throughput improvements and is preferable in class based storage assignment. The fact that decreasing the travel time by only 10% causes an economization of every tenth storage aisle with its S/R machine justifies the application of various strategies. This is achieved by 11 strategies in the miniload case and 18 strategies in the pallet case on average. As a result of our research, we conclude that implementing strategies composed of several elements of routing and sequencing should always be aimed for.

7.2 Conclusion and Outlook

After answering the research questions and summarizing our results, we encounter the boundaries of this thesis. There are questions we have left aside but consider them worth mentioning. We suggest further research to focus on the following aspects:

- How can the analytical results be extended to load handling devices with higher capacity or multiple-deep storage?
- Derivation of analytical expressions for strategies that select retrieval positions by sequencing of retrieval requests. The formulated models serve as a valuable starting point and has to be adjusted to characteristics of SKU selection. In this way, a multitude of the strategies included in the long list can be formulated.
- Investigation of the full potential of class based storage for quadruple command cycles within double deep storage environments. First, is there a better arrangement of zones with double deep storage, e.g. a separation into front and rear? And second, how much can throughput be increased by the strategies we have not implemented for class-based storage, e.g. Flip Flop combined with nearest neighbor storage and no-cost selection of retrievals or any kind of multiple storage?

Forecasts say, revenue in warehousing is constantly increasing both in Germany and the US (Statista 2016, Statista and US Census Bureau 2017) and revenues in e-commerce are expected to double between 2015 and 2020 (Digital Market Outlook 2016). Therefore, we expect the technology of AS/RSs to further evolve in the upcoming years, requiring academics to keep research up with the developments, such as multiple load handling systems or shuttle based S/RSs. Moreover, a number of questions arise when thinking about the progression of both AS/RSs and warehousing technology:

- Are there even more sophisticated policies to perform a quadruple command cycle we have not considered?

- To what extend can our results be transferred to further developed AS/RSs?
- How is control and efficient operation of AS/RSs achieved in fully connected environments of the fourth industrial revolution?

Especially the last question will produce a great number of research questions to be studied. The customers' increasing demand for availability and delivery time as well as dealing with a great amount of real-time data are only two examples of the challenges operators of warehouses face. This thesis makes a contribution by ensuring sufficient examination of existing systems and thus represents one important step for upcoming research.

Nomenclature

Basics

a_x	Acceleration and deceleration of the AS/RS in horizontal direction 38
a_y	Acceleration and deceleration of the AS/RS in vertical direction 38
b	Shape factor of a normalized, dimensionless rack according to Bozer and White (1984) 39
$E(DC)$	Expected travel time of a dual command cycle 47
$E(DC)^N$	Dimensionless expected travel time of a dual command cycle 45
$E(SC)$	Expected travel time of a single command cycle 44
$E(SC)^N$	Dimensionless expected travel time of a single command cycle 44
$E(SW_1)$	Expected one-way travel time from the I/O position to one randomly chosen position 42
$E(SW_m)$	Expected smallest one-way travel time between the I/O point and one of m randomly selected locations 54
$E(TB_1)$	Expected travel time between two randomly chosen positions 45
$E(TB_m)$	Expected travel time between one randomly selected position and the nearest of m randomly selected positions 53

$E(UF)$	Approximated nearest rearrangement distance for the discrete rack model according to Lippolt (2003)	57
H	Maximum lifting height of the AS/RS within the rack	37
L	Maximum travel distance of the AS/RS within the rack	37
T	Scaling factor for the normalized, dimensionless rack	39
t_a	Time-wise impact of acceleration and deceleration for one travel movement	38
t_{dead}	Dead times for reaction, control and operation of sensors	28
t_{mast}	Mast damping time per movement of the S/R machine	28
$t_{x,max}$	Maximum travel time in horizontal direction	39
$t_{y,max}$	Maximum travel time in vertical direction	39
v_x	Maximum speed of the AS/RS in horizontal direction	34
v_y	Maximum speed of the AS/RS in vertical direction	34
w	Shape factor according to Gudehus (1972b)	37

Assumptions and Analytical Models

$E(QC)^N$	Dimensionless expected travel time of a quadruple command cycle in single deep	70
$E(QC)_{dd}$	Expected travel time of a quadruple command cycle in double deep storage	89
$E(QC_{dd})_{Mod.Tango}$	Approximated travel time for the QC with the modified tango	100
$E(RC)$	Expected mean cycle time of a regular rearrangement command	70

$E(RD)$	Expected rearrangement distance to the nearest available position according to Lippolt (2003)	85
$E(t^R_{LHD})$	Mean load handling time for retrieval operations	88
$E(t^S_{LHD})$	Mean load handling time for storage operations	88
l	Total number of storage lanes	64
l^*	Total number of storage positions	64
$P(i)$	Probability that a storage lane is in state $i, i \in \{E, H, F\}$	72
$P(SRSR)$	Probability that the QC is executed according to the SRSR order	68
$P(SSRR)$	Probability that the QC is executed according to the SSRR order	68
pR	Number of potential storage lanes that are available for rearrangement	85
$R1 - R2$	Retrieval events in the general model	74
$S1 - S2$	Storage events in the general model	74
t_0	Component of the travel time allowing for dwell times	89
t_d	Time needed positioning the other LHD in front of the same storage lane	86
$t_{LHD,f}$	Access time of the LHD to the front position	66
$t_{LHD,r}$	Access time of the LHD to the rear position	66
$t_{LHD,tango}$	Path-depending time needed to perform a tango rearrangement	86
$t_{Tango,mod.}$	Total time needed to perform a modified tango rearrangement	100
t_{Tango}	Total time needed to perform a tango rearrangement	70
z	Filling level of the storage space	68
E	Empty state of a storage lane	65

F	Filled state of a storage lane 65
H	Half-filled state of a storage lane 65
nl	Number of storage lanes, used to measure the rearrangement distance 107

Strategy Definition

$\hat{P}(MS.Sim)$	Probability that simultaneous multiple storage is performed regarding the generating probability 164
$\hat{P}(MS.Suc)$	Probability that successive multiple storage is performed regarding the generating probability 166
$E(QC^{3S})^N$	Dimensionless expected travel time of the three-stop QC 143
$E(t_{LHD}^{S^{FF}})$	Mean load handling time of a storage operation that occurs during the Flip Flop 155
$E(TB_m^o)$	Expected o^{th} smallest travel-between time amongst one point and m randomly selected locations 143
$Nl_{no-cost}$	Mean number of storage lanes within the no-cost zone 159
$P(GP.Sim)$	Generating probability for a new pair of storage lanes that offers the possibility for multiple storage 164
$P(GP.Suc)$	Generating probability that a storage lane becomes empty and allows successive multiple storage 165
$P(MS.Sim)$	Probability that simultaneous multiple storage is performed 161
$P(MS.Suc)$	Probability that successive multiple storage is performed 164
$P(SL)$	Probability that there is an empty position within the no-cost zone 157

References

Arnold, D. and K. Furmans (2009). *Materialfluss in Logistiksystemen* (6., erw. Aufl. ed.). VDI. Berlin: Springer.

Arnold, D., H. Isermann, A. Kuhn, H. Tempelmeier and K. Furmans (2008). *Handbuch Logistik*. Berlin, Heidelberg: Springer Berlin Heidelberg.

Ascheuer, N., M. Grötschel and A. A.-A. Abdel-Hamid (1998). *Order picking in an automatic warehouse: Solving online asymmetric TSPs*, Volume SC 98,08 of *Preprint / Konrad-Zuse-Zentrum für Informationstechnik Berlin*. Berlin: ZIB.

Ashayeri, J., R. M. Heuts, M. Valkenburg, H. C. Veraart and M. R. Wilhelm (2002). A geometrical approach to computing expected cycle times for zonebased storage layouts in AS/RS. *International Journal of Production Research 40*(17), p. 4467–4483.

Atz, T. (2016). *Eine algorithmenbasierte Methode zur ganzheitlichen Systemplanung automatischer Hochregallager*. Dissertation, Technische Universität München.

Bartholdi, J. J. and S. T. Hackman (2016). *Warehouse & Distribution Science* (Release 0.97 ed.). Atlanta, Ga.: The Supply Chain and Logistics Institute.

Bayes and Price (1763). An Essay towards Solving a Problem in the Doctrine of Chances. By the Late Rev. Mr. Bayes, F. R. S. Communicated by Mr. Price, in a Letter to John Canton, A. M. F. R. S. *Philosophical Transactions of the Royal Society of London 53*(0), p. 370–418.

Bortolini, M., R. Accorsi, M. Gamberi, R. Manzini and A. Regattieri (2015). Optimal design of AS/RS storage systems with three-class-based assignment strategy under single and dual command operations. *The*

International Journal of Advanced Manufacturing Technology 79(9-12), p. 1747–1759.

Bozer, Y. A., E. C. Schorn and G. P. Sharp (1990). Geometric Approaches to Solve the Chebyshev Traveling Salesman Problem. *IIE transactions 22*(3), p. 238–254.

Bozer, Y. A. and J. A. White (1984). Travel-time models for automated storage/retrieval systems. *IIE transactions 16*(4), p. 329–338.

Braun, M. S. A. (2016). *Entwicklung, Analyse und Evaluation von Modellen zur Ermittlung des Energiebedarfs von Regalbediengeräten.* Dissertation, KIT, Karlsruhe.

Brunk, F. (2016). *Entwicklung von Spielzeitmodellen für Mehrfacheinlagerung und Shortest-Leg in doppeltiefen Lagern.* Master thesis supervised by K. Doerr, KIT, Karlsruhe.

Bühl, A. (2016). *SPSS 23: Einführung in die moderne Datenanalyse* (15., aktualisierte Auflage ed.). Hallbergmoos: Pearson Deutschland GmbH.

Chang, D. T., U.-P. Wen and J. T. Lin (1995). The impact of acceleration/deceleration on travel-time models for automated storage/retrieval systems. *IIE transactions 27*(1), p. 108–111.

Dambach Lagersysteme GmbH & Co. KG (15.10.2015). Kenndaten eines typischen RBG, Palette: E-Mail.

Devore, J. L. and K. N. Berk (2012). *Modern Mathematical Statistics with Applications.* New York, NY: Springer New York.

Digital Market Outlook (2016). Prognose der Umsätze im E-Commerce weltweit in den Jahren 2015 bis 2021 (in Milliarden Euro): according to de.statista.com. `https://de.statista.com/statistik/daten/studie/484763/umfrage/prognose-der-umsaetze-im-e-commerce-weltweit/`. (Accessed on 2017/07/21).

Domschke, W. (2007). *Logistik: Transport: Grundlagen, lineare Transport- und Umladeprobleme* (5., 5., überarbeitete Aufl. ed.). Oldenbourgs Lehr-

und Handbücher der Wirtschafts- u. Sozialwissenschaften. München: De Gruyter.

Dörr, K. and K. Furmans (2016a). Determination of Cycle Times for Double Deep Storage Systems Using a Dual Capacity Handling Device: Edited by Kimberly Ellis, Andres Carrano, Rene de Koster, Gary Forger, J., David Porter, and Jeffrey Smith. *Progress in Material Handling Research: 2016 14.*

Dörr, K. and K. Furmans (2016b). Durchsatzbetrachtungen für doppeltiefe Lager unter dem Einsatz von zwei Lastaufnahmemitteln. *Logistics Journal: Proceedings 2192*(9084), p. 313–326.

Epp, M. (2017). *Performance evaluation of shuttle-based storage and retrieval systems using discrete-time queueing network models.* Dissertation, KIT, Karlsruhe.

Eynan, A. and M. J. Rosenblatt (1993). An interleaving policy in automated storage/retrieval systems. *International Journal of Production Research 31*(1), p. 1–18.

Eynan, A. and M. J. Rosenblatt (1994). Establishing zones in single-command class-based rectangular AS/RS. *IIE transactions 26*(1), p. 38–46.

FEM 9.851 (1978). *Leistungsnachweis für Regalbediengeräte: Zuverlässigkeit, Verfügbarkeit.* Fédèration Européenne de la Manutention.

Gagliardi, J.-P., J. Renaud and A. Ruiz (2012). Models for automated storage and retrieval systems: A literature review. *International Journal of Production Research 50*(24), p. 7110–7125.

Garlock, P. (1997). *Berechnung der Umschlagleistung von Regalförderzeugen mit parallel ablaufenden Positionierungen der Lastaufnahmemittelachsen.* Dissertation, Technische Universität Graz, Graz.

Gebhardt Fördertechnik Gmbh (2017). Gebhardt Cheetah Miniload - Type 701. http://www.gebhardt-foerdertechnik.de/en/products/warehouse-technology/storage-and-retieval-cranes/cheetah-miniload/?desktop= (Accessed on 2017/08/02).

Gebhardt Fördertechnik Gmbh (9.10.2015). Kenndaten typisches RBG, AKL: E-Mail.

Glass, M. (2009). *Schnellläuferstrategien in Lagern und Dynamische Zonierung.* Dissertation, Technische Universität Dresden, Dresden.

Grafe, W. (1997). Was leisten Mehrfachlastaufnahmemittel. *Hebezeuge und Fördermittel 37*(4), p. 127–129.

Graves, S. C., W. H. Hausman and L. B. Schwarz (1977). Storage-Retrieval Interleaving in Automatic Warehousing Systems. *Management Science 23*(9), p. 935–945.

Gu, J., M. Goetschalckx and L. F. McGinnis (2010). Research on warehouse design and performance evaluation: A comprehensive review. *European Journal of Operational Research 203*(3), p. 539–549.

Gudehus, T. (1972a). Analyse des Schnelläufereffektes in Hochrellagern. *f+ h Fördern und Heben 22*(2), p. 65–66.

Gudehus, T. (1972b). Auswahlregeln für Geschwindigkeiten von Regalförderzeugen. *f+ h Fördern und Heben 22*(3), p. 33–35.

Gudehus, T. (1972c). Fahrwegoptimierung in automatisierten Hochregallagern. *f+ h Fördern und Heben 22*(3), p. 123–124.

Gudehus, T. (1972d). Grundlagen der Spielzeitberechnung für automatische Hochregallager. *Deutsche Hebe-und Fördertechnik 18*, p. 63–68.

Gudehus, T. (1972e). Regalförderzeuge für mehrere Ladeeinheiten. *f+ h Fördern und Heben 22*(11), p. 607–609.

Gudehus, T. and K. H. Hofmann (1973). Die optimale Höhe von Hochregallägern. *Deutsche Hebe-und Fördertechnik 2*, p. 68–72.

Guttman, I., S. S. Wilks and J. S. Hunter (1982). *Introductory Engineering Statistics* (3. ed. ed.). Wiley series in probability and mathematical statistics. New York: Wiley.

Han, M.-H., L. F. McGinnis, J. S. Shieh and J. A. White (1987). On sequencing retrievals in an automated storage/retrieval system. *IIE transactions 19*(1), p. 56–66.

Happes, M., K. Dörr and K. Furmans (2017). Determination of cycle times for operating policies in double deep storage systems using a dual capacity load handling device: Working Paper.

Hausman, W. H., L. B. Schwarz and S. C. Graves (1976). Optimal storage assignment in automatic warehousing systems. *Management Science: Journal of the Institute for Operations Research and the Management Sciences 22*(6), p. 629–638.

Hwang, H. and S. B. Lee (1990). Travel-time models considering the operating characteristics of the storage and retrieval machine. *International Journal of Production Research 28*(10), p. 1779–1789.

Johnson, M. E. and M. L. Brandeau (1996). Stochastic modeling for automated material handling system design and control. *Transportation Science 30*(4), p. 330–350.

Kalyanaraman, p. and C. Keerthika (2016). A Review on Automated Storage/ Retrieval Systems and Shuttle Based Storage/Retrieval Systems. *International Journal on Recent and Innovation Trends in Computing and Communication 4*(11), p. 167–171.

Kayser, J. (2003). *Spielzeiten doppeltiefer Hochregallager mit Mehrfachlastaufnahmemittel.* Diploma thesis supervised by Christian Lippolt, Universität Karlsruhe, Karlsruhe.

Keserla, A. and B. A. Peters (1994). Analysis of dual-shuttle automated storage/retrieval systems. *Journal of Manufacturing Systems 13*(6), p. 424–434.

Kim, J. and A. Seidmann (1989). *A framework for the exact evaluation of expected cycle times in automated storage systems with full-turnover item allocation and random service requests,* Volume WP 89/5 of *Quantitative methods.* Rochester: University of Rochester. Graduate School of Management.

Kohn, W. and R. Öztürk (2011). *Statistik für Ökonomen.* Springer-Lehrbuch. Berlin, Heidelberg: Springer Berlin Heidelberg.

Kouvelis, P. and V. Papanicolau (1995). Expected travel time and optimal boundary formulas for a two-class-based automated storage/retrieval system. *International Journal of Production Research 33*(10), p. 2889–2905.

Kraul, R. (2010). *Ersatzmodelle für die Leistungsbewertung von automatischen Lagersystemen.* Dissertation, Technische Universität München, München.

Lee, H. F. and S. K. Schaeffer (1996). Retrieval sequencing for unit-load automated storage and retrieval systems with multiple openings. *International Journal of Production Research 34*(10), p. 2943–2962.

Lee, H. F. and S. K. Schaeffer (1997). Sequencing methods for automated storage and retrieval systems with dedicated storage. *Computers and Indistrial Engineering 32*(2), p. 351–362.

Lerher, T., M. Šraml, I. Potrč and T. Tollazzi (2010). Travel time models for double-deep automated storage and retrieval systems. *International Journal of Production Research 48*(10/12), p. 3151–3172.

Lippolt, C. R. (2003). *Spielzeiten in Hochregallagern mit doppeltiefer Lagerung.* Dissertation, Universität Karlsruhe, Karlsruhe.

Martin, H. (2014). *Transport- und Lagerlogistik.* Wiesbaden: Springer Fachmedien Wiesbaden.

Mecalux, S. (2017a). Stacker crane for boxes. https://www.mecalux.com/automated-warehouses-for-boxes/stacker-cranes-for-boxes. (Accessed on 2017/08/02).

Mecalux, S. (2017b). Stacker cranes for pallets. https://www.mecalux.com/automated-warehouses-for-pallets/stacker-cranes-for-pallets. (Accessed on 2017/08/02).

Meller, R. D. and A. Mungwattana (1997). Multi-shuttle automated storage/retrieval systems. *IIE transactions 29*(10), p. 925–938.

MIAS Group (2017). News, Telescopic Forks and Stacker Cranes, AS/RS, transport technology. http://www.mias-group.com/en/. (Accessed on 2017/08/02).

Oser, J. and M. Ritonja (2004). Expected cycle time in a class-based single- and double-deep storage system. *Progress in Material Handling Research 8*, p. 310–325.

Potrč, I., T. Lerher, J. Kramberger and M. Šraml (2004). Simulation model of multi-shuttle automated storage and retrieval systems. *Journal of Materials Processing Technology 157-158*, p. 236–244.

Ritonja, M. (2003). *Spielzeiberechnungen und bedienstrategien von Regalbdiengeräten mit Mehrfach-Lastaufnahmemitteln.* Dissertation, Technische Universität Graz, Graz.

Roodbergen, K.-J. and I. Vis (2009). A survey of literature on automated storage and retrieval systems. *European Journal of Operational Research 194*(2), p. 343–362.

Rosenblatt, M. J. and A. Eynan (1989). Note—Deriving the Optimal Boundaries for Class-Based Automatic Storage/Retrieval Systems. *Management Science 35*(12), p. 1519–1524.

Rouwenhorst, B., B. Reuter, V. Stockrahm, G. J. van Houtum, R. J. Mantel and W. Zijm (2000). Warehouse design and control: Framework and literature review. *European Journal of Operational Research 122*(3), p. 515–533.

Rushton, A. (2017). *The Handbook of Logistics and Distribution Management: Understanding the Supply Chain* (6th ed. ed.). London: Kogan Page.

Sarker, B. R. and P. S. Babu (1995). Travel time models in automated storage/retrieval systems: A critical review. *International Journal of Production Economics 40*(2-3), p. 173–184.

Sarker, B. R., L. Mann and J. R. Leal Dos Santos (1994). Evaluation of a class-based storage scheduling technique applied to dual-shuttle automated storage and retrieval systems. *Production Planning & Control 5*(5), p. 442–449.

Sarker, B. R., A. Sabapathy, A. M. Lal and M.-H. Han (1991). Performance evaluation of a double shuttle automated storage and retrieval system. *Production Planning & Control 2*(3), p. 207–213.

Schaab, W. (1969). *Automatisierte Hochregalanlagen: Bemessung und Wirtschaftlichkeit.* Dissertation, Technische Universitaet Berlin, Berlin.

Schwarz, L. B., S. C. Graves and W. H. Hausman (1978). Scheduling Policies for Automatic Warehousing Systems: Simulation Results. *A I I E Transactions 10*(3), p. 260–270.

Seemüller, S. (2006). *Durchsatzberechnung automatischer Kleinteilelager im Umfeld des elektronischen Handels.* Dissertation, Technische Universität München, München.

SSI Schäfer Fritz Schäfer Gmbh (2017). Exyz storage-retrieval machine. https://www.ssi-schaefer.com/de-de/produkte/lagern/palettenlagerung/regalbediengeraet-exyz-9142. (Accessed on 2017/08/02).

Statista (2016). Prognostizierte Umsatzentwicklung in der Lagerei und Verkehrsdienstleistungsbranche in Deutschland in den Jahren von 2007 bis 2020 (in Milliarden Euro): in Logistik- und Verkehrsdienstleistungen Branchenreport, according to de.statista.com. `https://de.statista.com/statistik/daten/studie/248626/umfrage/prognose-zum-umsatz-in-der-lagerei-und-verkehrsdienstleistungsbranche/`. (Accessed on 2017/07/21).

Statista and US Census Bureau (2017). Umsatz der Branche Allgemeine Lagerhaltung in den USA von 2008 bis 2015 und Prognose bis zum Jahr 2020 (in Millionen Euro): according to de.statista.com. https://de.statista.com/prognosen/423937/allgemeine-lagerhaltung-in-den-usa—umsatzprognose. (Accessed on 2017/07/21).

ten Hompel, M. and T. Schmidt (2010). *Warehouse Management: Organisation und Steuerung von Lager- und Kommissioniersystemen* (4. Aufl. ed.). VDI-Buch. s.l.: Springer-Verlag.

Van den Berg, J. P. (1996). Class-based storage allocation in a single-command warehouse with space requirement constraints. *International Journal of Industrial Engineering 3*(1), p. 21–28.

Van den Berg, J. P. (1999). A literature survey on planning and control of warehousing systems. *IIE transactions 31*(8), p. 751–762.

Van den Berg, J. P. and A. Gademann (1999). *Optimal routing in an automated storage.* Estados Unidos: Institute of Industrial Engineers.

Vasili, M. R., S. H. Tang and M. Vasili (2012). Automated storage and retrieval systems: a review on travel time models and control policies. In: *Warehousing in global supply chain : advanced, models, tools and applications for storage systems,* p. 159–209. London [u.a.]: Springer.

VDI Richtlinie 3651 (1973). *Testspiele zum Leistungsvergleich und zur Abnahme von Regalförderzeugen.* Düsseldorf: VDI-Verlag.

VDI Richtlinie 4480, Blatt 4 (1998). *Durchsatz von automatischen Lagern mit mehrfachtiefer Lagerung.* Düsseldorf: VDI-Verlag.

Waldmann, K.-H. and U. M. Stocker (2013). *Stochastische Modelle: Eine anwendungsorientierte Einführung* (2., Aufl. ed.). EMILA-stat. Berlin: Springer.

Wen, U. P., D. T. Chang and S. P. Chen (2001). The impact of acceleration/deceleration on travel-time models in class-based automated S/R systems. *IIE transactions 33*(7), p. 599–608.

Xu, X., G. Shen, Yugang Yu and W. Huang (2015). Travel time analysis for the double-deep dual-shuttle AS/RS. *International Journal of Production Research 53*(3), p. 757–773.

Zörnig, P. (2016). *Probability Theory and Statistical Applications: A Profound Treatise for Self-Study.* De Gruyter Textbook. Berlin, Boston: De Gruyter.

Zschau, U. (1963). *Technisch-wirtschaftliche Studie über die Anwendbarkeit von Stapelkranen im Lagerbetrieb.* Dissertation, Technische Universitaet Berlin, Berlin.

List of Figures

List of Tables

A Additional Analytics

In the following additional information with regard to the analytical models presented in Chapter 4 and 5 are provided.

Limit value observation for $P(SSRR) \to 0$

We examine the behavior of $P(E), P(H)$ and $P(F)$ (see equations 4.29, 4.30 and 4.31) when $P(SSRR) \to 0$. Therefore, let $x = P(SSRR)$. Unknown is:

$$\lim_{x \to 0^+} P(E)$$

Let:

$$\frac{2z + xz - \sqrt{x^2 z^2 - 12xz^2 + 4xz + 4z^2 + 8z + 4} + 2}{2x} =: \frac{f(x)}{g(x)}$$

$f(x)$ and $g(x)$ are continuously differentiable in $\mathbb{R}$ having:

$$f'(x) = z - \frac{1}{2} \frac{2xz^2 - 12z^2 + 4z}{\sqrt{x^2 z^2 - 12xz^2 + 4xz + 4z^2 + 8z + 4}}$$

$$g'(x) = 2$$

And

$$\lim_{x \to 0^+} f'(x) = f'(0) = z - \frac{1}{2} \frac{-12z^2 + 4z}{\sqrt{4z^2 + 8z + 4}}$$
$$= z - \frac{-2z(3z-1)}{\sqrt{(2z+2)^2}} = \frac{z(2z+2) + 2z(3z-1)}{2z+2}$$
$$= \frac{2z^2 + 2z + 6z^2 - 2z}{2z+2} = \frac{4z^2}{z+1}$$

$$\lim_{x \to 0^+} g'(x) = g'(0) = 2$$

It is further valid that $\lim_{x \to 0^+} f(x) = \lim_{x \to 0^+} g(x) = 0$.

L'Hôspital's rule implies that:

$$\lim_{x \to 0^+} \frac{f(x)}{g(x)} = \lim_{x \to 0^+} \frac{f'(x)}{g'(x)} = \frac{2z^2}{z+1}$$

It follows that:

$$\lim_{x \to 0^+} P(E) = \lim_{x \to 0^+} (\frac{f(x)}{g(x)} - 2z + 1)$$
$$= (\lim_{x \to 0^+} \frac{f(x)}{g(x)}) - 2z + 1$$
$$= \frac{2z^2}{z+1} - 2z + 1$$
$$= \frac{2z^2 - 2z^2 - 2z + z + 1}{z+1}$$
$$= \frac{1-z}{1+z} \qquad (q.e.d.)$$

Which is in accordance to the result of Lippolt (2003, p.134). In the same way, we can show that:

$$\lim_{x\to 0^+} P(H) = \frac{2z(1-z)}{1+z}$$

$$\lim_{x\to 0^+} P(F) = \frac{2z^2}{1+z}$$

Probabilities from the modified tango model

The changed probabilities of R3, R4 and R5 in the variation of the basic model for modified tango operation:

$$\begin{aligned} P(R3) = & \tfrac{P(F)}{P(H)+2P(F)} \cdot [\tfrac{3}{4}\tfrac{P(H)}{P(E)+P(H)} \\ & +(1-\tfrac{1}{2}P(SSRR)) \cdot (1-P(F)^2) \cdot \tfrac{P(H)}{P(E)+P(H)}] \end{aligned} \tag{A.1}$$

$$\begin{aligned} P(R4) = & \tfrac{P(F)}{P(H)+2P(F)} \cdot [\tfrac{3}{4}\tfrac{P(E)}{P(E)+P(H)} \\ & +(1-\tfrac{1}{2}P(SSRR)) \cdot (1-P(F)^2) \tfrac{P(E)}{P(E)+P(H)}] \end{aligned} \tag{A.2}$$

$$P(R5) = \tfrac{P(F)}{P(H)+2P(F)} \cdot \tfrac{1}{2}P(SSRR) \cdot P(F)^2 \tag{A.3}$$

$(\frac{1}{3}, \frac{2}{3})$ Model

The probability distribution of the execution order is $P(SSRR, SRSR) = \frac{1}{3}, \frac{2}{3}$. The results are:

- Stationary storage lane allocation

$$P(E) = 4 + \frac{3}{2}z - \frac{1}{2}\sqrt{z^2 + 84z + 36} \tag{A.4}$$

$$P(H) = 6 - 5z + \sqrt{z^2 + 84z + 36} \tag{A.5}$$

$$P(F) = 3 + \frac{7}{2}z - \frac{1}{2}\sqrt{z^2 + 84z + 36} \tag{A.6}$$

- Rearrangement probability

$$P_{Rearrange} = \frac{4\,(6+7z-\sqrt{z^2+84z+36})}{z} \tag{A.7}$$

$$P(Regular) = \frac{5\,(6+7z-\sqrt{z^2+84z+36})}{24\,z} \tag{A.8}$$

$$P(Tango) = \frac{6+7z-\sqrt{z^2+84z+36}}{24\,z} \tag{A.9}$$

- Rearrangement distance

$$E(RD) = -2-\frac{7}{2}z+\frac{1}{2}\sqrt{z^2+84z+36} \tag{A.10}$$

- Mean load handling times

$$\begin{aligned} E(t^S_{LHD}) =& \frac{1}{2}\cdot(t_{LHD,f}\cdot(1+\frac{-6-5z+\sqrt{z^2+84z+36}}{-2-\frac{7}{2}z+\frac{1}{2}\sqrt{z^2+84z+36}}) \\ &+ t_{LHD,r}\cdot(1-\frac{-6-5z+\sqrt{z^2+84z+36}}{-2-\frac{7}{2}z+\frac{1}{2}\sqrt{z^2+84z+36}})) \end{aligned} \tag{A.11}$$

-

$$\begin{aligned} E(t^R_{LHD}) =& \frac{1}{2}\cdot(t_{LHD,f}\cdot(\frac{6+11z+\sqrt{z^2+84z+36}}{4z}) \\ &+ t_{LHD,r}\cdot(\frac{-3-3z+\sqrt{z^2+84z+36}}{4z}) \end{aligned} \tag{A.12}$$

Using the terms stated above in the following equation (A.13), we obtain the cycle time for a quadruple command cycle without tango.

$$\begin{aligned} E(QC_{dd}) =& t_0+\frac{5}{2}\cdot(\frac{v_x}{a_x}+\frac{v_y}{a_y})+E(QC)^N\cdot\frac{L}{v_x} \\ &+\frac{5}{6}\cdot P_{Rearrange}\cdot E(RC)+\frac{1}{6}\cdot P_{Rearrange}\cdot t_{Tango} \\ &+4\cdot E(t^S_{LHD})+4\cdot E(t^R_{LHD})-2\cdot t_{LHD,f} \end{aligned} \tag{A.13}$$

Detailed Validation Results

The validation results from chapter 4.5 can be found in the TableA.1 and A.2.

Reference value	Analytical Model	Simulation Model	95% confidence interval		Delta rel.
			lower	upper	
$E(QC_{dd})_{SSRR}$	63.0454 s	62.9181 s	62.8883 s	62.9478 s	0.202%
$P_{Rearrangement}$	0.2331	0.2335	0.2328	0.2341	−0.157%
P_{Tango}	0.2331	0.2339	0.2333	0.2346	−0.357%
% Storage at front	0.7562	0.7530	0.7524	0.7537	0.442%
% Storage at rear	0.2438	0.2470	0.2463	0.2476	−1.308%
% Retrieval from front	0.4662	0.4661	0.4653	0.4671	0.014%
% Retrieval from rear	0.5338	0.5339	0.5329	0.5348	−0.013%
P(E)	0.0392	0.0395	0.0394	0.0395	−0.664%
P(H)	0.1216	0.1211	0.1209	0.1212	0.428%
P(F)	0.8392	0.8395	0.8394	0.8395	−0.031%
Mean rearr. distance	1.3155	1.2722	1.2711	1.2733	3.292%
Mean rearr. time	12.4367 s	11.5220 s	11.5209 s	11.5230 s	7.355%

Table A.1: Validation of the analytical SSRR model: Computed vs. simulated results.

The two right-hand columns in Tables A.1 and A.2 show that the approximation for the modified tango model does not lead to a substantial loss of accuracy. The deviation of the mean travel time lies in between the results of the basic model and the SSRR model. However, the influence of the approximation is reflected by the state probabilities of the storage lanes as they show greater deviations compared to the former cases. Thus, also the proportions of storage and retrieval at front and rear position show greater differences. This is because the state probabilities for the modified tango model are approximated using those of the general model (see section 4.4). The probability that a storage lane is empty, $P(E)$, differs more than 6%, while the probabilities $P(H)$ and $P(F)$ more than 4%. For the previously presented models, deviations are less than 1%.

Reference value	Analytical Model	Simulation Model	95% confidence interval		Delta rel.
			lower	upper	
$E(QC_{dd})_{TangoOpt.}$	63.2824 s	63.0542 s	63.0318 s	63.0766 s	0.361%
$P_{Rearrangement}$	0.3528	0.3544	0.3537	0.3551	−0.461%
P_{Tango}	0.1176	0.1182	0.1176	0.1189	−0.529%
$P(Tango)_{standard}$	0.0843	0.0854	0.0850	0.0858	−1.343%
$P(Tango)_{modified}$	0.0333	0.0328	0.0325	0.0331	1.521%
% Storage at front	0.6954	0.6749	0.6742	0.6755	2.955%
% Storage at rear	0.3046	0.3251	0.3245	0.3258	−6.746%
% Retrieval from front	0.4704	0.4712	0.4704	0.4729	0.174%
% Retrieval from rear	0.5296	0.5288	0.5280	0.5296	−0.155%
P(E)	0.0467	0.0489	0.0488	0.0490	−6.746%
P(H)	0.1066	0.1022	0.1021	0.1024	4.703%
P(F)	0.8467	0.8489	0.8488	0.8490	−4.121%
Mean rearr. distance	1.3396	1.2860	1.2850	1.2869	4.001%
Mean rearr. time	12.5071 s	11.6079 s	11.6062 s	11.6095 s	7.190%

Table A.2: Validation of the analytical model with modified tango: Computed vs. simulated results.

Compatibility of Strategy Elements

	Additional rules							Selection of single requests			
Main routing policy	No Cost	Simple NN	RNN	Storage near retrieval	Retrieval near storage	Minimum Perimeter	Avoided/ forced rearrangement	Fixed storage positions	Free storage selection	Fixed retrieval requests	Free retrieval selection
Shortest Leg	x		x		x				+	o	o
Simple NN			x					+	o	o	o
RNN		x						+	o	o	o
Storage near retrieval			x						+	o	o
Retrieval near storage			x					o	o		+
Flip Flop	x	x	x	x		x			+	o	o
Multiple Storage	x	x	x	x	x		x		+	o	o
Increase Tango	x		x		x		x	o	o	o	o
Class based Storage								o	o		+

x Combination possible

\+ combination required
o combination possible

Figure A.1: Overview of the possible combination of design elements for the strategy definition

Analytical determination of NN1

The determination of the exact order of NN1 is illustrated by the tree diagram shown in Figure A.2.

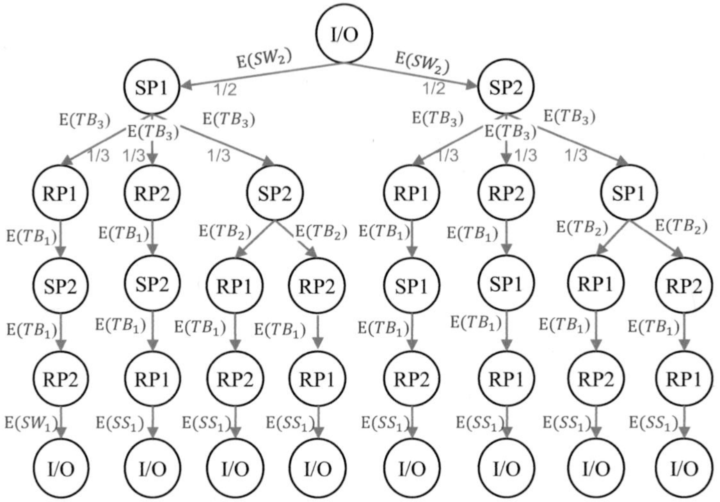

Figure A.2: Simple nearest neighbor policy (NN1): Procedure of selection and cycle determination

Analytical determination of NN3

The determination of the exact order of NN3 is illustrated by the tree diagram shown in Figure A.3. The dotted lines emphasize that the order within the retrieval positions is already defined at the beginning, however they are approached last in the cycle. Consequently, the gray lines represent the actual order of the storage and retrieval jobs with the corresponding probabilities for every branch.

Derivation of $E(TB_m)$ and $E(TB_m^o)$

$E(TB_m)$ and $E(TB_m^o)$ can be derived by means of order statistics.

$E(TB_m)$ is the normalized mean travel between distance from one randomly selected point (x, y) to the nearest of m randomly selected points in the $(1xb)$ rack model. The distribution function $Q(\zeta)$ and the density function $q(\zeta)$ of the normalized travel between among two randomly selected points are known from equations 3.48 and 3.49. The normalized travel be-

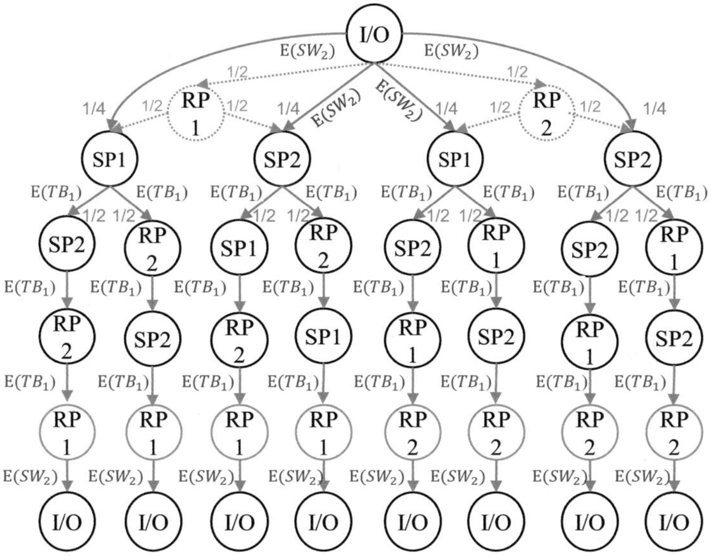

Figure A.3: Revers nearest neighbor policy (NN2): Procedure of selection and cycle determination

tween distance among one random point (x, y) and one of m randomly selected points $(x_i, y_i), i \in \{1, 2, ..., m\}$ is described by the same functions. Let ζ_i be the normalized travel between distances from (x, y) to (x_i, y_i). We are looking for the minimum travel between distance among m travel between distances which is (see equation 3.1):

$$TB_m = min\{\zeta_1, \zeta_2, \cdots, \zeta_m\} = \zeta_{(1)} \tag{A.14}$$

By means of the cdf of the first order statistics, the probability that TB_m is less or equal $\zeta \in [0, 1]$, for any shape factor b, is:

$$Q_{Z(1)}(\zeta) = 1 - (1 - Q(\zeta))^m \tag{A.15}$$

With the density function:

$$q_{Z(1)}(\zeta) = mq(\zeta)(1 - Q(\zeta)^{m-1} \tag{A.16}$$

Using equations 3.48 and 3.49 and setting up the expected value, we can derive from A.16:

$$E(TB_m) = \int_0^1 \zeta mq(\zeta)(1 - Q(\zeta))^{m-1} d\zeta \tag{A.17}$$

$E(TB_m^o)$ is the normalized mean travel between distance from one randomly selected point (x, y) to the $o^{th.}$-nearest of m randomly selected points in the $(1xb)$ rack model. Again, the distribution function $Q(\zeta)$ and the density function $q(\zeta)$ from equations 3.48 and 3.49 apply here. We are looking the $o^{th.}$ smallest travel between distance from m distances $\zeta_1, \zeta_2, ..., \zeta_m$ which is (see equation 3.1):

$$TB_m^o = \zeta_{(o)} \tag{A.18}$$

By means of the cumulative distribution function of the $i. - th$ order statistics the probability that TB_m^o is less or equal $\zeta \in [0, 1]$, for any shape factor b, with $i = o$ and $n = m$ is:

$$Q_{Z(o)}(\zeta) = \sum_{l=o}^{m} \binom{m}{l} Q(\zeta)^l (1 - Q(\zeta))^{(m-l)} \tag{A.19}$$

With the density function:

$$q_{Z(o)}(\zeta) = \frac{m!}{(o-1)!(m-o)!} q(\zeta) Q(\zeta)^{(o-1)} (1 - Q(\zeta))^{(m-o)} \tag{A.20}$$

Using equations 3.48 and 3.49 and setting up the expected value, we can derive from A.20:

$$E(TB_m^o) = \int_0^1 \zeta \frac{m!}{(o-1)!(m-o)!} q(\zeta) Q(\zeta)^{(o-1)} (1 - Q(\zeta))^{(m-o)} d\zeta \tag{A.21}$$

The left-hand side (A) and the right-hand side (B) of Figure A.4 show examples for the meaning of $E(TB_m)$ and $E(TB_m^o)$, respectively.

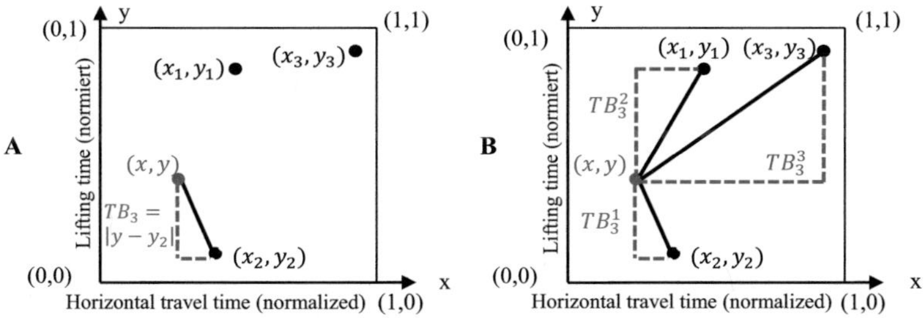

Figure A.4: Example of TB_3 (A) and TB_3^1, TB_3^2 and TB_3^3 (B) in the normalized rack model

Validation of MS1

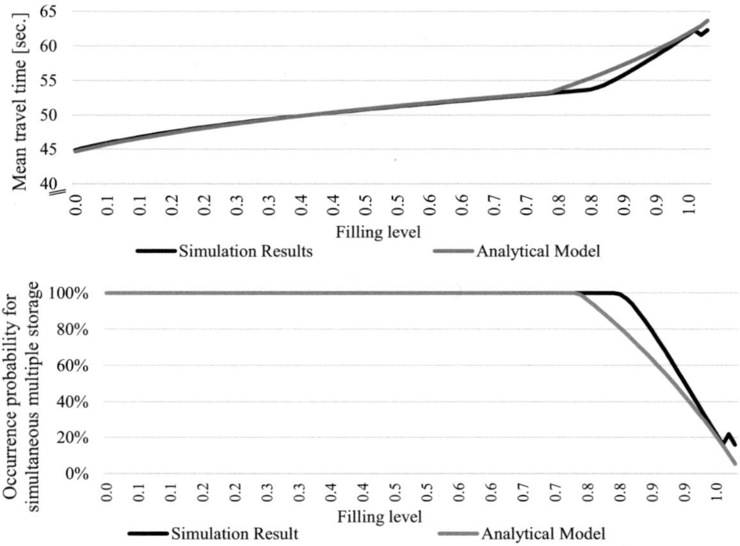

Figure A.5: Comparison of simulation and analytical model for simultaneous multiple storage

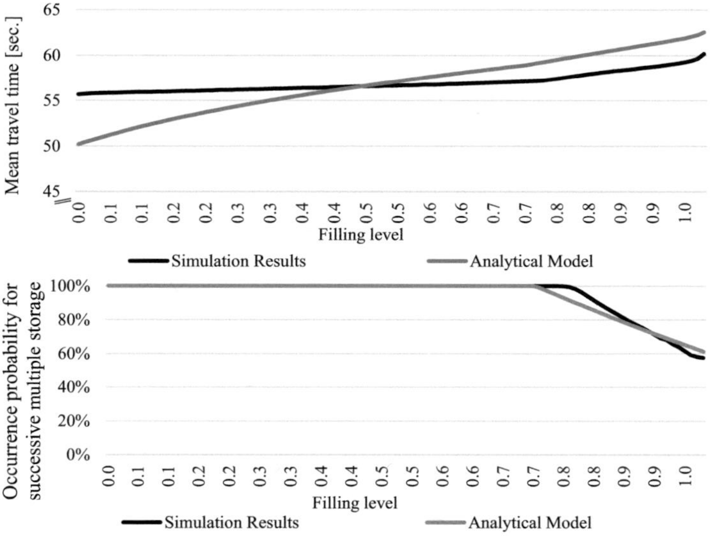

Figure A.6: Comparison of simulation and analytical model for successive multiple storage

Parameter evaluation for a scaling factor of 40

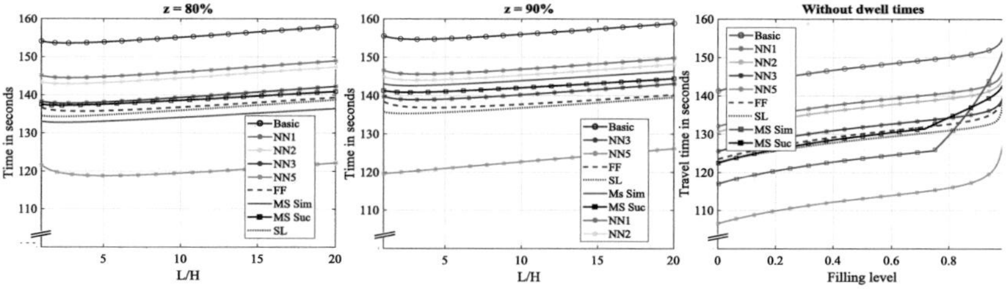

Figure A.7: Travel time evaluations with $L/v_x = 40$.

B Simulation Studies

Warm-up Phase

As shown in Chapter 4, the allocation of the storage lanes is no random allocation but dependent on the way the storage place is operated, i.e. the amount of tango that is applied. The simulation model is set up with a randomly filled storage rack which means the specific storage lane allocation is only reached after a certain amount of cycles. In order to analyze the influence of the warm-up phase we do evaluate two different cases. First, we compare the simulation results in the regular randomly filled set-up. In the second case, we start the simulation with an unevenly filling of the rack, i.e. the storage lanes are filled in a row from the bottom to the top of the rack. Based on the graphical representation of the mean cycle time over time, the length of the war-up phase is determined. In the first case the length of the warm-up phase is determined to 1000 cycles for the second case, we allow a warm-up phase of 2000 cycles.

Reference value	Randomly filled	Unevenly filled
Mean travel time of 100,000 cycles	63.0916 s	63.0880 s
Mean travel time without the warm-up phase	63.0910 s	63.0834 s
Difference	0.0006 s	0.0046 s

Table B.1: Evaluation of the influence of the warm-up phase on simulation results. The warm-up is 1000 cycles for the randomly filled rack and 2000 cycles for the unevenly filled rack.

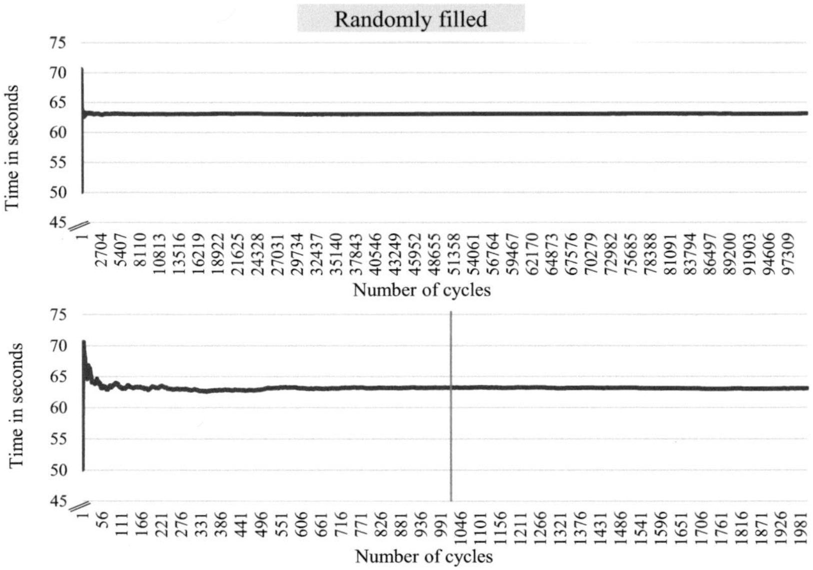

Figure B.1: Development of the mean cycle time over time for all cycles (top) and the first 2000 (bottom) when randomly filled at the beginning. At the bottom the warm-up phase of 1000 cycles is marked with the vertical line.

Table B.1 shows the results of this evaluation for which we used the unmodified validation set-up. In the randomly filled rack, the difference when cropping the warm-up phase is 0.0006 seconds, while for the unevenly filled rack the difference is 0.0046 seconds.

Additional Results of Simulation Studies

In the following, Figure B.3 shows the cumulative distribution function of the three SKU profiles. Every profile consists of 312 different SKU types. The Figures B.4, B.5 and B.6 illustrate the influence of the retrieval options as discussed subsection 6.3.2 The Figures B.7, B.8 and B.9 illustrate the influence of the filling level as discussed subsection 6.3.3.

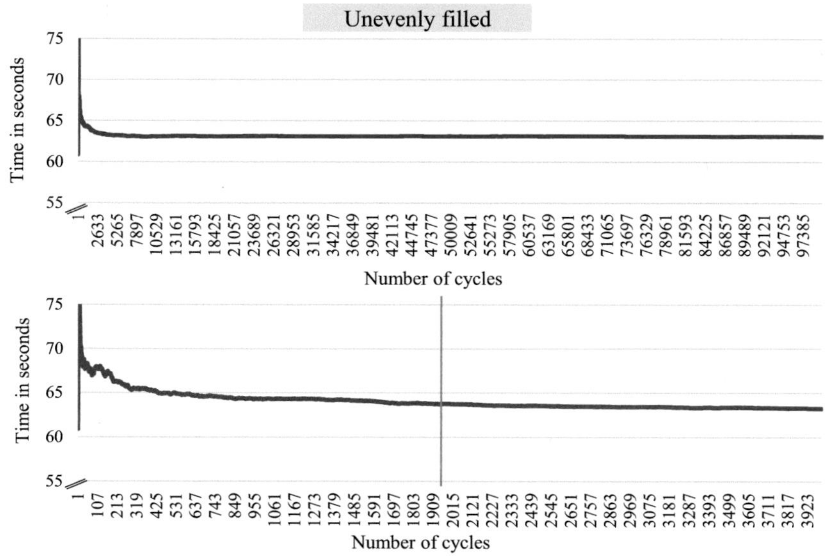

Figure B.2: Development of the mean cycle time over time for all cycles (top) and the first 4000 (bottom) when unevenly filled at the beginning. At the bottom the warm-up phase of 2000 cycles is marked with the vertical line.

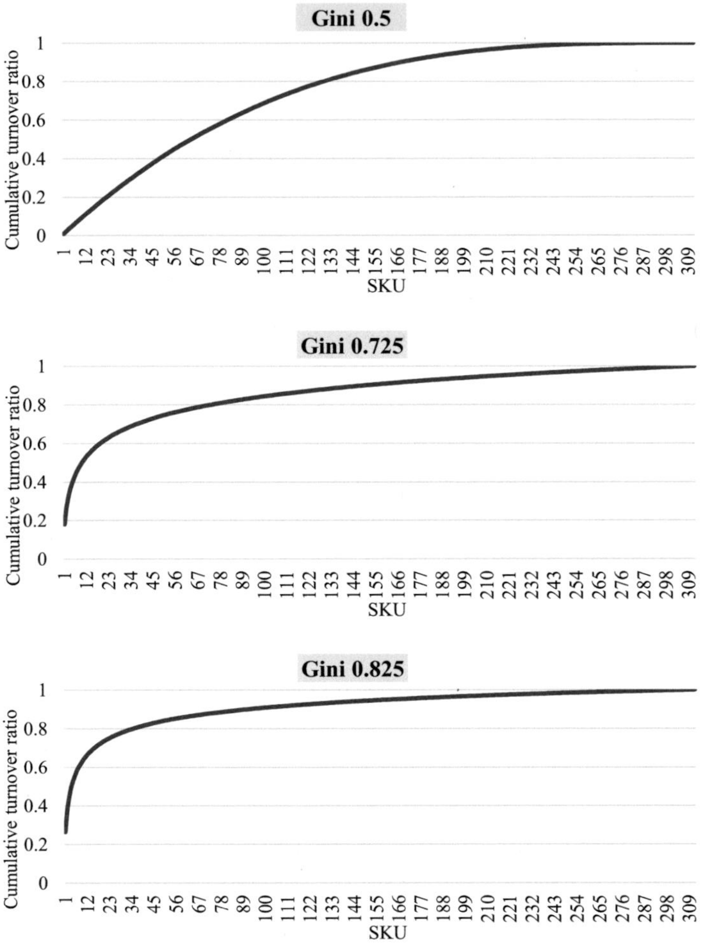

Figure B.3: Cumulative distribution function of the SKU's demand probability

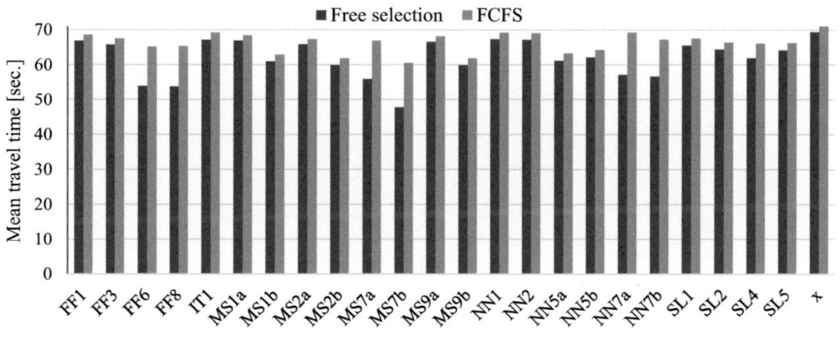

Figure B.4: Comparison of simulation results for different retrieval options - Miniload scenario with random storage assignment

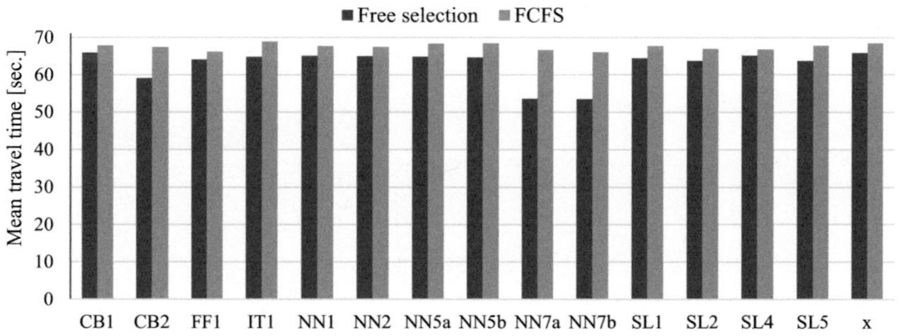

Figure B.5: Comparison of simulation results for different retrieval options - Miniload scenario with class based storage assignment

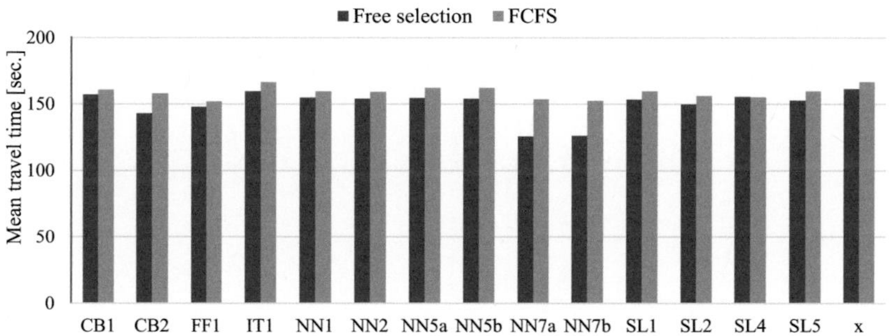

Figure B.6: Comparison of simulation results for different retrieval options - Pallet scenario load scenario with class based storage assignment

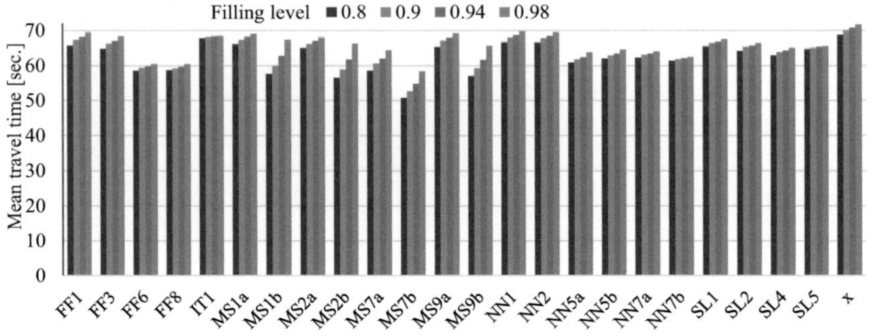

Figure B.7: Comparison of simulation results for different filling levels in the miniload scenario with random storage assignment

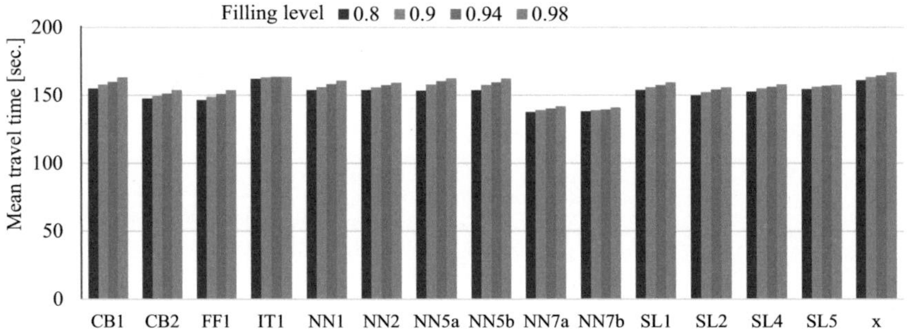

Figure B.8: Comparison of simulation results for different filling levels in the pallet scenario with class based storage assignment

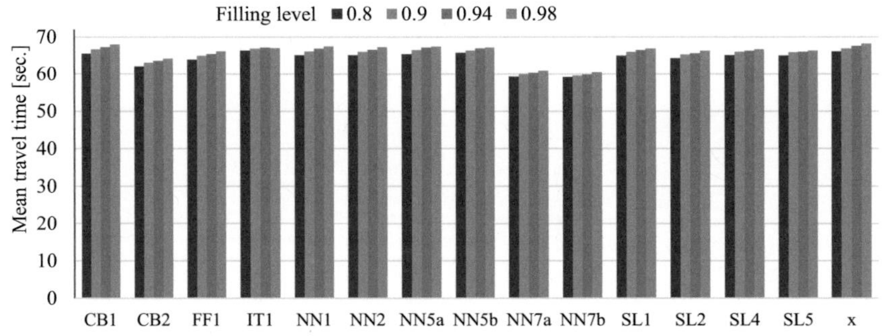

Figure B.9: Comparison of simulation results for different filling levels in the miniload scenario with class based storage assignment

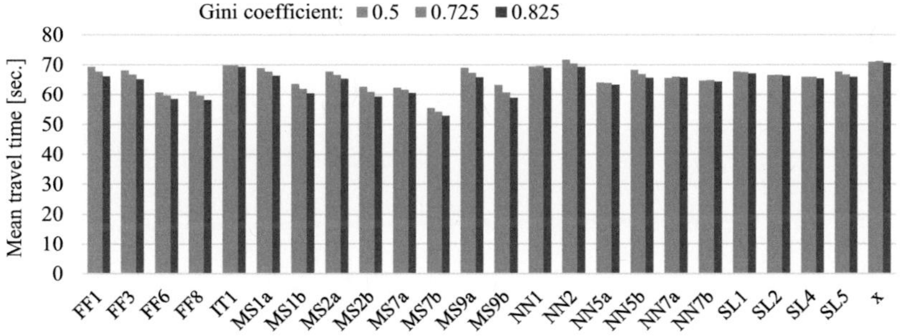

Figure B.10: Comparison of simulation results for different Gini coefficients of the SKU distribution in the miniload scenario with random storage assignment

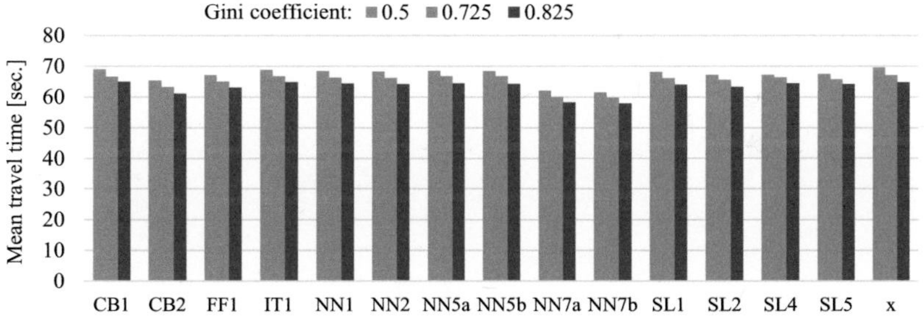

Figure B.11: Comparison of simulation results for different Gini coefficients of the SKU distribution in the miniload scenario class based storage assignment